Disruptive Logic Architectures and Technologies

Pierre-Emmanuel Gaillardon
Ian O'Connor · Fabien Clermidy

Disruptive Logic Architectures and Technologies

From Device to System Level

Springer

Pierre-Emmanuel Gaillardon
INF 339 (Bâtiment INF)
EPFL IC ISIM LSI1
Station 14
1015 Lausanne
Switzerland

Fabien Clermidy
CEA-LETI
MINATEC
Rue des Martyrs 17
38054 Grenoble
France

Ian O'Connor
Bâtiment F7
INL, Site Ecole Centrale de Lyon
36 Avenue Guy Collongue
69134 Ecully
France

ISBN 978-1-4899-9231-4 ISBN 978-1-4614-3058-2 (eBook)
DOI 10.1007/978-1-4614-3058-2
Springer New York Heidelberg Dordrecht London

Softcover reprint of the hardcover 1st edition 2012

Printed on acid-free paper

Springer is part of Springer Science+Business Media (www.springer.com)

Preface

For over four decades, the semiconductor industry has experienced exponential growth. According to the *International Technology Roadmap for Semiconductors*, as we advance deeper into the era of nanotechnology, traditional *Complementary Metal Oxide Semiconductor* electronics will soon reach its physical and economical limits. This book deals with the opportunities offered by disruptive technologies and in particular explores novel designs for digital architectures (with an emphasis on reconfigurable structures).

In a first approach, we will base our investigations around conventional FPGA architectures, and examine ways in which disruptive technologies can lead to structural improvements. In a more prospective approach, we will then also present some novel architectural schemes for ultra-fine grain computing, where the inherent properties of the considered disruptive technologies lead to reduced size of the logic elements.

All these improvements are explored in a unified way by a new methodological approach, which allows the fast evaluation of an emerging technology on a prospective architecture. We use this to discuss their impact, from the level of simple circuits to that of complex architectures.

Several technologies, ranging from 3D integration of devices (Phase Change Memories, Monolithic 3D, Vertical NanoWire-based transistors) to dense 2D arrangements (Double-Gate Carbon Nanotubes, Sublithographic Nanowires, Lithographic Crossbar arrangements), have been envisaged. Novel architectural organizations, as well as the associated tools, are presented in order to explore this largely uncharted design space.

This book represents a new step in the field of nanoarchitectures. It makes the link between technology and applications by proposing prospective designs and toolflows adapted to the requirements of both sides. It thus lends precious insight into the field of emerging technologies for designers as well as technologists. Indeed, this book considers disruptive technologies from an architectural perspective, while all technological assumptions are validated. Thus, this book covers a broad range of fields that must be addressed holistically in such prospective research.

Contents

Acronyms

1-D	One Dimension
2-D	Two Dimensions
3-D	Three Dimensions
5T-SRAM	Five Transistors Static Random Access Memory
ALD	Atomic Layer Deposition
ANN	Artificial Neural Network
ASIC	Application Specific Integrated Circuit
BEC	Bottom EleCtrode
BEOL	Back-End-Of-Line
BJT	Bipolar Junction Transistor
BLE	Basic Logic Element
BLIF	Berkeley Logic Interchange Format
BTB	Band-To-Band
CAD	Computer Aided Design
CB	Connection Box
CB-NWFET	CrossBar of NanoWires Field Effect Transistors
CBRAM	Conductive-Bridging Random Access Memory
CLB	Configurable Logic Block
CMOL	CMos/nanOeLectronic
CMOS	Complementary Metal-Oxide-Semiconductor
CN	Carbon Nanotube
CNLB	Configurable "Nano" Logic Block
CNT	Carbon NanoTube
CNTFET	Carbon NanoTube Field Effect Transistor
CVD	Chemical-Vapor-Deposition
DG-CNFET	Double-Gate Carbon Nanotube Field Effect Transistor
DIBL	Drain-Induced Barrier Lowering
DRAM	Dynamic Random Access Memory
DSP	Digital Signal Processor/Processing
DUV	Deep UltraViolet
EDA	Electronic Design Automation

EPFL	Ecole Polytechnique Fédérale de Lausanne
ERD	Emerging Research Devices
ERM	Emerging Research Materials
EUV	Extreme UltraViolet
EV	EValuation
FDSOI	Fully Depleted Silicon-On-Insulator
FEOL	Front-End-Of-Line
FET	Field Effect Transistor
FF	Flip-Flop
FGLC	Fine Grain Logic Cell
FO4	FanOut-of-four
FPGA	Field Programmable Gate Array
GAA	Gate-All-Around
GIDL	Gate-Induced Drain Leakage
GNR	GrapheneNanoRibbon
GST	$Ge_2Sb_2Te_5$
HVT	High Threshold Voltage
IC	Integrated Circuit
IMS	Laboratoire d'Intégration du Matériau au Système
INL	Institut des Nanotechnologies de Lyon
IP	Intellectual Property
ITRS	International Technology Roadmap for Semiconductor
LB	Logic Block
LETI	Laboratoire d'Electronique et de Technologie de l'Information
LP	Low Power
LUT	Look-Up Table
LVT	Low Threshold Voltage
MAC	Multiplier Accumulator
MCluster	Matrix Cluster
MCNC	Microelectronics Center of North Carolina
MIN	Multistage Interconnection Network
MOS	Metal-Oxide-Semiconductor
MOSFET	Metal-Oxide-Semiconductor Field Effect Transistor
MPPA	Massively-Parallel Processor Array
MPSOC	Multi-Processor System-On-Chip
MPW	Multi Project Wafer
MRAM	Magnetic Random Access Memory
MUX	MUltipleXer
NASIC	Nanoscale Application Specific Integrated Circuit
NEMS	Nano-Electro Mechanical System
NOC	Network-On-Chip
NVM	Non-Volatile Memory
NW	NanoWire
NWFET	NanoWire Field Effect Transistor
OxRAM	Oxide Random Access Memory

PAL	Programmable Array Logic
PC	Phase-Change
PC	PreCharge
PCM	Phase-Change Memory
PDP	Power-Delay Profile
PIDS	Process Integration, Devices and Structures
PLA	Programmable Logic Array
PLAD	PLAsma Doping
PSP	State University-Philips
RAM	Random Access Memory
RR	Routing Resource
RRAM	Resistive Random Access Memory
ReRAM	Resistive Random Access Memory
SB	Switch Box
SCE	Short-Channel Effects
SNOW	Silicon Nanoelectronics On 300-mm Wafer
SiNW	Silicon NanoWires
SOC	System-On-Chip
SOI	Silicon-On-Insulator
SRAM	Static Random Access Memory
T-VPACK	Timing-driven Versatile PACKer
TCAD	Technology Computer Aided Design
TEC	Top Electrode
TSV	Through-Silicon-Vias
UFF	Unbalanced Flip-Flop
ULSI	Ultra Large Scale Integration
VLSI	Very Large Scale Integration
VPACK	Versatile PACKer
VPR	Versatile Place and Route tool
VTR	Verilog-To-Routing suite
WKB	Wentzel-Kramers-Brillouin

Chapter 1
Introduction

Abstract In this chapter, we aim to introduce the global context of the book. While the microelectronics industry is still lead by the scaling, we point out its limits and highlights some novel way coming from the nanotechnologies. In order to provide an efficient evaluation strategy of the different ways, we present the global methodology used in the remaining of the book.

Our post-industrial way of life can be defined by an ever-growing need for mass consumer products. Every day, industry generates novel innovations that continually boost the market. Planes are increasingly automated, cars assist the drivers in a various number of situations for security or leisure (e.g. speed control, parking assistance, etc.), cell phones are able to communicate over the internet and video game consoles are controlled by handset motions or even player movements. Such evolutions mean that the systems have to be more intelligent, have simpler interfaces and be more communicative. The enabler behind all these novelties is the generalized use of embedded electronic systems, which have long become an integral and pervasive part of the human society. As the demand in terms of entertainment, data volume and connectivity is in constant growth, the system performance needs to improve more every day. The market is boosted by application requirements and remains in constant progress and evolution.

Electronic systems are complex systems, which aim to address computation in a specific target application. They are built around a physical hardware core, which can be programmed by loading and adding software modules. The hardware is the basic key for computation, in the sense that it is used to manage information at the physical level, i.e. in conventional systems and at the most elementary level, the electronic charge. The hardware is closely related to design and manufacturing techniques. The design defines the organization of the components in the system (i.e. the architecture). The manufacturing techniques define the physical components. This makes the related field highly multi disciplinary.

The basic unit of electronic systems is the transistor. Invented in 1947 by William Shockley, John Bardeen and Walter Brattain, the transistor is able to work

P.-E. Gaillardon et al., *Disruptive Logic Architectures and Technologies*,
DOI: 10.1007/978-1-4614-3058-2_1,

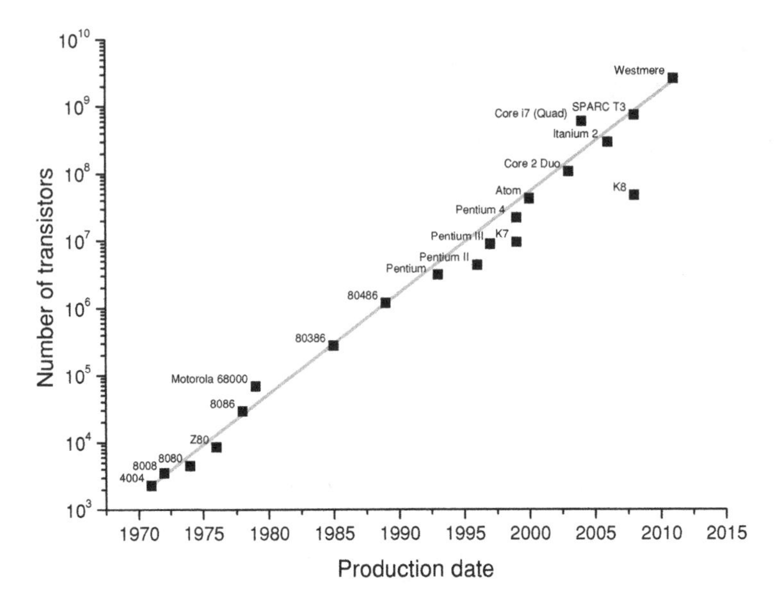

Fig. 1.1 Processor transistor count and Moore's law since 1971

as a switch at the smallest scale [1]. While the first demonstrated transistor was a bipolar transistor made of Germanium, the technology moved quickly to Silicon-based transistor and to the demonstration of the first *Metal-Oxide-Semiconductor Field Effect Transistor* (MOSFET) in 1959 [2]. The Silicon MOSFET transistor is still in use in the most advanced modern integrated circuits. By scaling the device and using a large number of transistors on a same substrate, it became possible to create integrated circuits of higher complexity and functionality. The era of the semiconductor industry has been defined by scaling, and consequent improvement for almost half a century: the industry has followed an exponential evolution from the first realization of integrated circuits to the current *Very Large Scale Integration* (VLSI)[1] with billions of transistors.

1.1 The Scaling Era

The way in which the semiconductor industry has evolved is unique with the scope of industrial history by its impact on society and rate of progress. Since the creation of the transistor, the semiconductor industry has grown exponentially, a trend which has been enshrined in Moore's law [3] (Fig. 1.1). Starting in 1965 and

[1] This term was first coined at the 100 k transistor mark. Other terms were later proposed (e.g. ULSI) to reflect the continually increasing complexity of integrated circuits , but remains the term of choice within the community.

Table 1.1 Linear scaling rules impact on device parameters [4]

Parameters	Scaling factor
Transistor length and width (L, W)	$1/\alpha$
Junction depth (x_j)	$1/\alpha$
Oxide thickness (t_{ox})	$1/\alpha$
Doping concentration (N_d, N_a)	α
Supply voltage (V_D)	$1/\alpha$
Drive current (I_D)	$1/\alpha$
Electric field (E)	1
Capacitance ($\varepsilon.A/t_{ox}$)	$1/\alpha$
Delay time ($\tau = C.V_D/I_D$)	$1/\alpha$
Power dissipation ($\sim V_D.I_D$)	$1/\alpha^2$
Device density ($\sim$ 1/A)	α^2

based on empirical observation, this law predicts an exponential growth (a doubling) of the number of transistors per die for each new technology generation. Even the growth rate (or time between technology generations) has not been constant though during recent decades, it still has a factor of two very 2 years. As illustrated by Fig. 1.1, it is worth noticing that the transistor count of processors has followed a similar trend for the last four decades.

The increase in the number of transistors is an obvious consequence of scaling. Scaling leads to the reduction of the dimensions of elementary drawn transistors, making it is thus possible to integrate more devices within the same area. Further scaling brings many further advantages additional to that of density. The majority of technological parameters are scaled according to a given scaling strategy. The impact of scaling on the main device parameters for constant electric field (the dominant scaling strategy for deep submicron technology generations) is presented in Table 1.1. It is important to observe that scaling leads to a reduction of device power by a factor of α^2 and a speed up in intrinsic delay by a factor of α.

This fast evolution is unique. As an illustration, if other industries had followed a comparable rate of change, a one-way ticket between Paris and New York may would now cost only 0.01€, while the total flight time would have been reduced down to 0.25 s. Similarly, a car scaled by the same proportions would only cost 20 €, reach a speed of 3,000 km s^{-1}, would only weigh 10 mg and would only consume 1 l of fuel per 100,000 km.

First only seen as a trend based on empirical observation, Moore's law has become much more than just a speculation. It has been regarded (and is still considered) as a real specification for growth, enabling the definition of objectives for research and development in the semiconductor industry. This has given rise in recent years to the *International Technology Roadmap for Semiconductors* (ITRS), compiled by a consortium of academic and industrial leaders in the field of semiconductors, and whose goal is to survey the trends in semiconductor technology and predict its future evolution up to 15 years ahead. As a consequence, it defines the objectives for the next decades, forecasts the trend for future technology nodes below 10 nm gate length and identifies roadblocks to overcome.

1.2 New Challenges

The ITRS has indicated significant future hurdles that may impede the original scaling law. Limitations are found at various levels, from fundamental device characteristics to advanced system design methodologies.

1.2.1 Process Integration and Devices

The process integration and the devices will suffer from leakage and quantum effects, as well as from intrinsic fabrication hurdles (for example lithography steps).

1.2.1.1 Electrostatic Channel Control

With channel scaling, *Short Channel Effects* (SCE) are becoming increasingly dominant. The short channel effect reflects the lowering of the threshold voltage with a decreasing channel length. This effect is due to a two-dimensional distribution of the surface potential in the channel. Furthermore, in short channel devices, the *Drain-Induced Barrier Lowering* (DIBL) effect is also present. The DIBL consists of making the threshold voltage dependent on the drain bias. Both of these effects lower the threshold voltage and make the devices more vulnerable to variability.

1.2.1.2 Leakage

It is worth noticing that when the lateral dimensions of a transistor are scaled, the oxide thickness is also scaled. This leads to an exponential growth of the tunnel current through the gate oxide, thereby an increase in the gate leakage.

Moreover, new physical phenomena appear at scaled dimensions and make a significant to leakage current. Indeed, high values of channel doping cause *Band-To-Band* (BTB) tunneling and *Gate-Induced Drain Leakage* (GIDL). These effects lead to higher power consumption [5, 6].

1.2.1.3 Lithography

Microelectronics fabrication is based on lithography. Tools using *Deep UltraViolet* (DUV) light with wavelengths of 248 and 193 nm allow minimum feature sizes down to 50 nm. However, scaling below this will require several innovations in terms of design and equipments. The mask design must contain a high degree of regularity, in order to minimize the fabrication variability and impact on lithography steps, while *Optical Proximity Correction* (OPC) is required to compensate inherent diffraction. Finally, new equipments are envisaged to reduce the wavelength down to *Extremely Ultra-Violet* (EUV).

1.2.2 Systems and Tools

On the design perspective, several obstacles must also be overcome. Indeed, keep growing complexity of systems requires researches to develop new computation paradigms and the associated architectures, while design tools and methodologies should be improved in order to reduce the inherent development costs.

1.2.2.1 System Design

Many device parameters do not scale exactly according to the scaling theory [7], due to intrinsic device hurdles. As a result, many device parameters which used to be fixed by the technology node are becoming design parameters. This is especially true for the supply and threshold voltages. The designer can optimize these parameters in order to obtain the best trade-off in terms of area, power and delay.

Furthermore, other circuit parameters do not follow the scaling rules at all. Indeed, the latency of on-chip wires does not follow the same trend as front-end devices. Global signals impact the performance metrics significantly, and therefore are very challenging. In order to both improve the energy-per-bit and to increase the bandwidth density, new on-chip connection paradigms are explored such as *Network-On-Chip* (NOC) [8, 9].

Another challenging issue is the distribution of clock signals. Conventional circuits are based on a single clock that is distributed throughout the whole circuit. With the increase of the relative length of clock lines (compared to the technology node), the task of distributing the clock signal without skew (i.e. simultaneously) and without jitter (i.e. periodically) everywhere on the chip is extremely difficult.

1.2.2.2 Design Tools and Methodologies

The possibilities enabled by current technologies are enormous. Nevertheless, the number of potential transistors that are available to chip design is much larger than the number of transistors (or complexity) that design teams can efficiently use. This difference between the technological possibilities and the design capabilities is called the Design Productivity Gap. In order to improve the productivity, design tools are developed to automate the design process, as much as possible. In this way, hardware description languages, high-level synthesis and hardware/software co-integration are a small illustration of the main design tasks that need to be addressed deeply to reach the full potential of a given technology.

1.3 Architectural and Technological Opportunities

The semiconductor industry has to face many issues to pursue Moore's law, which in any case is expected to end when the MOSFET transistor reaches its physical limits. In this book, we will assess a twofold opportunity. This opportunity is based on reconfigurable architectures that can be improved by disruptive technologies, in order to yield solutions for future architectures.

1.3.1 Reconfigurable Architectures

In order to keep following Moore's law and to achieve the computing capacities necessary for future software applications, it is today widely recognized that *Systems-On-Chip* (SoC) will move initially towards *Multi-Processor Systems-On-Chip* (MPSoC), then towards *reconfigurable platforms*. These systems will be used in the majority of solutions and in particular for high-performance computing (analysis and modeling of complex phenomena, advanced human–machine interaction) and for low-power mobile systems (sensor networks…).

The reconfigurable approach to computing systems offers several advantages. It allows volume manufacturing and thus constitutes a solution to the projected evolution of mask costs. The mask costs are expected to be above $100 M in 2018 [10]. In all probability, "full-custom" dies will disappear or migrate towards the development of *Multi Project Wafer* (MPW) lines for very high performance applications; the appearance of "stacked" circuits based on 3D integration will aim at heterogeneous applications; and the reconfigurable approach will enable volume applications [10]. Such systems can cover a broad range of applications, and their performance levels exceed very clearly those of programmable systems in terms of computing speed, while requiring only one set of masks. Moreover, the natural association of these architectures with fault-tolerant design techniques enables robust architectures to be built in the context of increasingly unreliable elementary nanometric MOS devices. Nevertheless, the various types of reconfigurable circuits (FPGA, coarse-grain reconfigurable systems) are at a disadvantage (compared to "full-custom" solutions) in terms of performance and device count necessary to fulfill a specific function (Fig. 1.2).

1.3.2 Technological Evolutions

In this context, the emergence of new devices offers the opportunity to provide novel building blocks, to elaborate non-conventional techniques for reconfigurable design and consequently to reconsider the paradigms of architecture design.

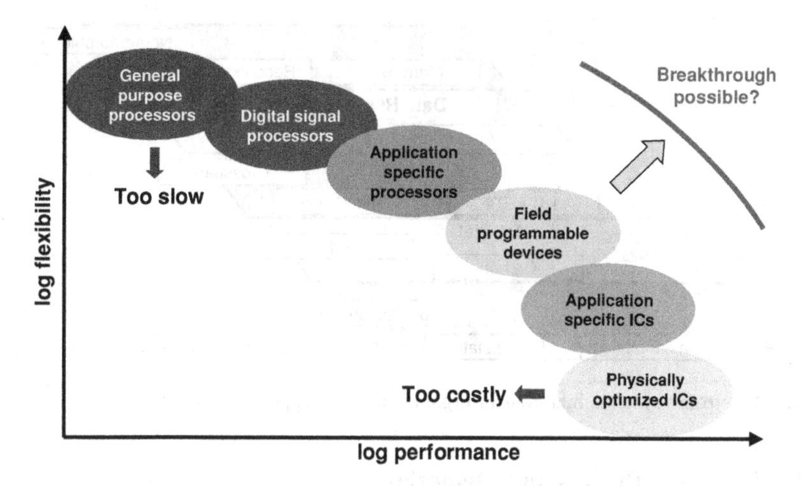

Fig. 1.2 Spectrum of implementation solutions for data processing algorithms

The ITRS *Emerging Research Device* (ERD) and *Emerging Research Materials* (ERM) chapters propose emerging technology fields for prospective research [11]. Two different directions are envisaged:

- the extension of the MOSFET device to other geometries and materials, and
- the use of other technologies and/or state variables for computing.

1.3.2.1 MOS Extensions

The straightforward evolution for MOSFET processes is to work on the channel materials and structure. The main idea consists of replacing the gate channel by new materials with high carrier mobility. Three different possibilities can be considered. The conventional approach consists of working on process engineering to improve the channel performance, with for example strained silicon or silicon-on-insulator wafers. The ultimate goal of such channel processing is to achieve the use of 1D structures as enhanced channels. Thus, silicon NanoWires could be seen as promising candidates for this evolution. Finally, the use of unconventional materials could be seen as a replacement material for silicon. Carbon electronics is a good prospective approach, with promising electrical properties, coupled with the confinement that is achievable by the structure. Furthermore, it is important to highlight that carbon nanotubes and graphene also exhibit respectively 1D and 2D structures that have superior electrical properties.

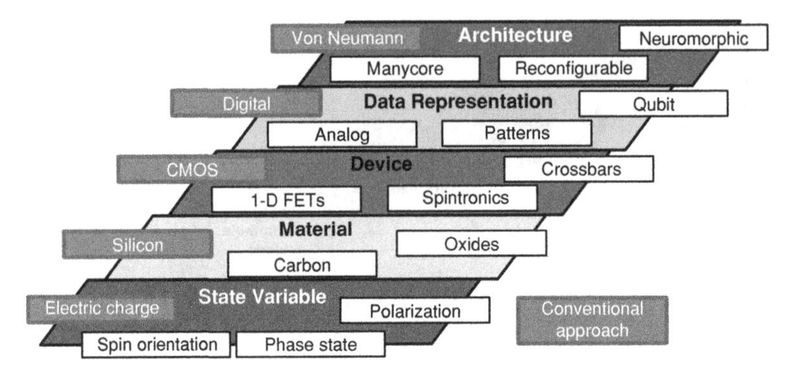

Fig. 1.3 Electronic systems hierarchical organization and opportunities [11]

1.3.2.2 Emerging Devices and Memories

Rather than considering "only" an improvement of the conventional electrical charge based MOSFET, the ITRS ERD and ERM chapters also suggest the possibility of using new state variables for computation as well as for the information storage. As shown in Fig. 1.3, even if the standard way to carry out computation is based on Silicon MOSFETs in Von Neumann multicore system architectures, many other ways could be explored. Indeed, unconventional state variables might be used, such as molecular state and phase state. It is also worth mentioning that unconventional solutions might be used at every hierarchy level, and that all these choices and/or combination of choices might give rise to new computation paradigms.

1.4 The Nanoscale Chicken and Egg Dilemma

While the use of emerging devices appears to be the correct approach to go beyond the limitations of planar CMOS technology, high cost and lack of maturity of fabrication processes render the task of finding a good candidate device difficult. In fact, the evaluation of any new technology is at the interface of different domains, and generally leads to a Chicken and Egg Dilemma, as depicted in Fig. 1.4.

Technology developments are driven by the application. As an illustration, we could oppose low power memories and high performance applications. These applications have different constraints regarding the speed, the power consumption and many other parameters, which lead to different CMOS optimization strategies, such as HVT[2] and LVT[3] transistors.

---

[2] High Threshold Voltage (V_t).

[3] Low Threshold Voltage (V_t).

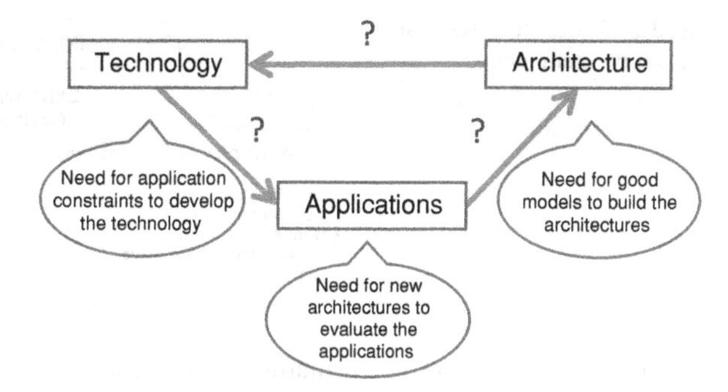

Fig. 1.4 The chicken and egg dilemma of disruptive technologies

Nevertheless, the formulation of an application constraint is performed through an architectural model. The model expresses the computational power offered by a technology, and allows to efficiently map algorithms onto the architecture.

Finally, the definition of an architecture requires information from the technology, in order to create an optimized system. It thus appears increasingly complicated to solve all the development pillars.

From this viewpoint, the following questions are raised:

- How can we find a suitable technology for future applications?
- Developing a technology generates excessive costs. It is thus impossible to carry out technological developments for all candidate processes. This indicates that we should employ a fast evaluation methodology for emerging technologies: But in this case, which architecture could we use? How will we evaluate the performance of a disruptive technology (i.e. which applications?) and which tools should we use?
- How can the technology characteristics be taken into account within the architecture? What should be the specific functions to address be? How can the computing blocks be improved? How can the peripheral circuitries be improved? What is the best arrangement of the logic blocks for efficient tuning of the architecture?

1.5 Objective and Methodology of the Book

Facing this broad range of questions, this book deals with the opportunities available from the use of disruptive technologies. It thus realizes the link between technology and applications by proposing prospective designs and toolflows in accordance with the requirements of both sides.

To enable scientists to perform the evaluation of a new technology from the architectural side, we propose a global evaluation methodology. The proposed methodology is presented in Fig. 1.5. A large set of available technologies will be evaluated under a generic architectural template. For the sake of genericity, the chosen template will be a generalized reconfigurable architecture. Depending on

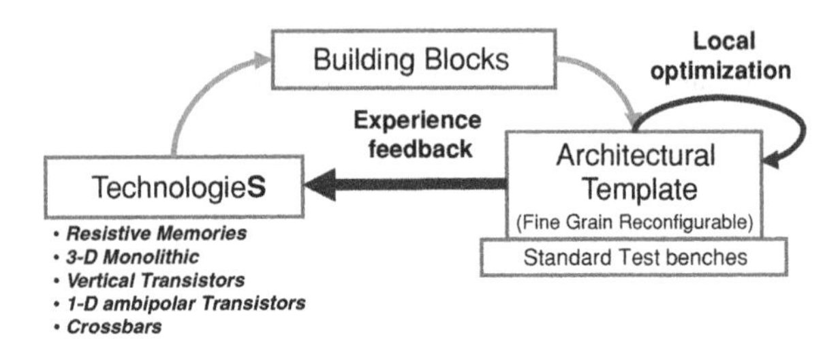

Fig. 1.5 Template-based fast evaluation methodology

the process and technological maturity, different models will be used and injected into the architectural template description. These models range from simple behavioral type to detailed models based on advanced physical *Technology Computer Aided Design* (TCAD) extractions and real-circuit measurement. The envisaged applications will be standard testbench sets in order to investigate a wide range of application domains and attempt to identify the best sets. A fast architectural optimization loop exists in the methodology. It allows identification of the best architectural trade-off for a given technology. Thus, it ensures that bad architectural sizing, which may lead to an erroneous estimation of technology performance, is avoided. Architectural evaluation results are finally the most important part of the complete methodology. Indeed, they are intended to feedback to the technology side, in order to find which parameters best improve the technology for the target application. Hence, this approach moved the problem to a global optimum, instead of considering only local optimizations.

1.6 Original Contributions of the Book

Most of the literature in the field offers solutions on very precise and focused questions. However, the emerging technology question is so broad that it is not possible to consider only a short part of the story. In this book, we will demonstrate how novel proposals originating from the technology/device level may directly impact complex system design. We expect the approach taken in this book to improve the work efficiency in the field, and to initiate more interactions between technologists and designers. Several questions have been assessed in different fields. Contributions have been made at several levels, ranging from elementary circuits to architectures, as well as benchmarking tools and methodologies.

1.6.1 Methodology and Tools

On the methodology and tools side, this book contributes to a state-of-the-art benchmarking toolset by proposing a complete tool flow suited to the design exploration of reconfigurable logic based on emerging devices (Chap. 6). Indeed, in order to evaluate and compare in a fair way any prospective circuit or architecture, it is mandatory to have a complete tool, able to instantiate a set of standard circuits based on the proposed technology and architecture. The proposed tool flow is based on extensions to the conventional tools from the reconfigurable community. The extensions allow a wide variety of structures and architectures for architectural exploration such as the management of logic cells instead of the conventional FPGA look-up table, the organization of logic cells into regular matrices and the use of fixed and incomplete interconnection topologies. To do so, a specific packing tool, called MPack, has been developed. It will be presented and its internal structure will be described in detail.

1.6.2 Memories and Routing Resources for FPGA

As we will discover in the next chapter, the FPGA architecture suffers from the amount of resources required for programmability. In particular, the versatility of the architecture is handled by several pass transistors connected to programming memories. These circuits are distributed throughout the whole circuit, and are known to be area- and power-hungry. In this book, we will assess three different technologies, in order to improve the memory and the routing aspects of reconfigurable systems (Chap. 3).

Resistive memories are promising devices which exhibit programmable non-volatile bistable resistive states. This technology allows a passive programmable device to be embedded into the metal layers. Thus, we will propose a configuration memory node. This node will use a voltage divider to store a logic level intrinsically. Then, the impact of the structure will be studied in comparison to flash memories. Using the favorable on-resistance property of resistive memory, we also propose to integrate them directly into the data path. This is obviously of high interest to create non-volatile and reprogrammable switches with low resistance in the on-state and low area impact. In addition, a resistive memory based switchbox will be proposed and compared to an equivalent state-of-the-art non-volatile implementation.

Instead of using only a passive device in the back-end layer, we will assess the opportunity to integrate active devices in a monolithic 3D arrangement. Monolithic 3D processes aim to stack several layers of active Fully-Depleted Silicon-On-Insulator. Since layers are processed sequentially, it is possible to obtain a high alignment quality, as well as a high via density between the two layers. We will propose an elementary FPGA block implementation using this 3D technology. Effectively, we will split the memory part from the data paths of the FPGA blocks. This will allow distinct optimization of the sizing and the processes.

Finally, we will push the 3D concept to the limits by proposing a "true" 3D implementation of routing resources. Vertical NanoWire Field Effect Transistors have been demonstrated as a possible technology to create vertical transistors between metal lines. These transistors are large, which enables them to reach promising performance levels. We then propose in building a "smart" back-end methodology with complex configurable vias, as well as configuration memories for vias and signal buffers. We will propose a performance evaluation methodology based on TCAD simulations to open the way towards prediction of the most advanced technologies.

1.6.3 Logic Blocks for FPGA

While improving memories and routing resources appears to be the most convenient way to improve the FPGA architecture, it is worth pointing out that the structure uses only a very small part of the computational real-estate. The largest part of the circuit is used for "peripheral circuitry". Thus, we could expect to find a more effective structure. In this way, it is interesting to work on logic blocks and see how emerging devices could lead to new paradigms for computing at fine-grain (Chap. 5). Two main device families will be studied.

Firstly, we will study a 1D ambipolar technology (here based on Double-Gate Carbon Nanotube Field Effect Transistor technology). These devices could be configured between *n*- and *p*-type conduction by changing the voltage of an extra polarity gate. This allows in-field reconfigurability to be achieved at the device level. We called this kind of device an "enhanced-functionality" device. This technology has been used to provide a very compact logic cell, able to perform computation at an ultra-fine level of granularity. We will refine the technological assumptions and generalize the proposal of the use of a configuration back-gate to a family of standard cells.

Secondly, we will focus on "density-increased" devices. Emerging technologies have opened the way towards the use of active devices in a high density crossbar fashion. We thus propose to use crossbars of nanowires to realize a compact ultra-fine grain logic cell. The proposed crossbar process involves an FDSOI-based lithographic process flow. We will explain how circuits and layouts could be implemented based on this technology.

These developments lead to significant reduction of logic block size and to the use of finer levels of granularity. In the rest of the book, we will refer to this by the term "ultra-fine grain logic".

1.6.4 Nanoscale Architectures

Ultra-fine grain logic provides a large reduction of logic real-estate. However, the scaling of wires and peripheral resources is not so direct. Hence, new organizations must be envisaged to assess an efficient architecture. We will propose regular arrangements of logic cells into matrices (Chap. 6). It should be noted that such an arrangement is also motivated by regular patterning in lithography steps or in

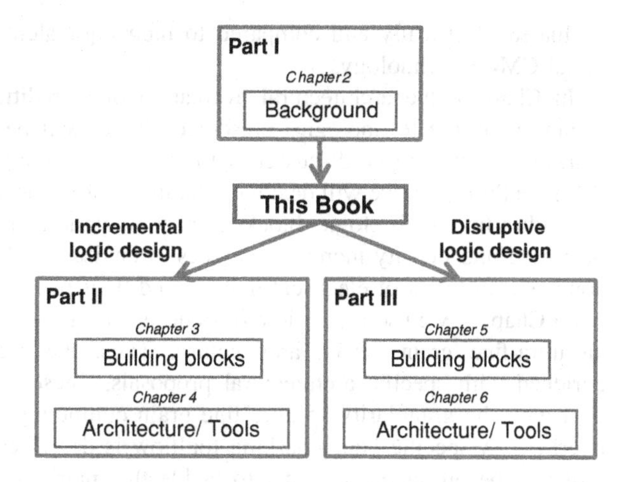

Fig. 1.6 Organization of the book

self-assembly. While the use of a standard FPGA interconnect scheme could lead to an unexpected overload in terms of resources, we propose a fixed and incomplete interconnect pattern. This pattern ensures maximum shuffling, reachable by scarce resource availability. This regular pattern is then used to realize logic blocks for reconfigurable architectures. A performance evaluation of the structure will be performed and it will be compared to standard FPGA. More disruptive architectures will then be discussed.

1.7 Organization of the Book

The book is composed of five main chapters organized into three parts, not including the introduction and the concluding chapter. The global organization of the book is depicted in Fig. 1.6.

In the first part, we describe the background and motivation of the work with a state-of-the-art Field Programmable Gate Array architecture, where tendencies and issues are described. Then, we will summarize existing computing architectures, based on emerging devices. We will particularly focus on post-CMOS heterogeneous architectures, as they appear to be the most relevant to our field. Finally, we will formulate and generalize the description of reconfigurable circuits in terms of hierarchical levels. We will also define the architectural template that will be the baseline of our architectural evaluation.

In the second part, we will target an incremental improvement of reconfigurable designs. Chapter 3 will focus on the use of emerging technologies for routing and memory implementations in reconfigurable circuits. Resistive Memory and Vertical transistor technologies, as well as Monolithic 3D integration processes, will be studied to assess the impact of moving the routing resources to the back-end layer. In this chapter, the performance of each proposed circuit will be evaluated electrically and compared to their equivalent counterparts in conventional CMOS technology.

In Chap. 4, the architectural evaluation of a traditional FPGA architecture, enhanced by the technologies presented above, will be proposed. Then, benchmarking results will be discussed with regards to the equivalent CMOS circuit.

In the third part, we will deal with disruptive logic design. Chapter 5 will assess new ultra-fine grain logic blocks for computation. The use of functionality-increased and density-increased devices is envisaged. The logic blocks are then compared in terms of electrical metrics and their overall potential is discussed.

In Chap. 6, we use the logic blocks designed in the preceding stage to develop an ultra-fine grain FPGA architecture. Thus, the reconfigurable template is enriched with specific architectural proposals. These proposals are expected to deal with the granularity of ultra-fine grain computing and with interconnection overload. A specific benchmarking toolflow is described. In particular, a packer tool will be presented, in order to tackle the matrices of cells with fixed interconnections. Finally, benchmarking results will be presented and discussed.

In Chap. 6, the book is concluded.

References

1. J. Bardeen, W.H. Brattain, Three-electrode circuit element utilizing semiconductor materials, US Patent No. 2524035 (1948)
2. D. Kahng, Electric field controlled semiconductor device, US Patent No. 2524035 (1960)
3. G.E. Moore, Cramming more components onto integrated circuits, Electronics **38**(8), 114–117 (1965)
4. H. Iwai, Roadmap for 22 nm and beyond (Invited Paper). Microelectron. Eng. **86**(7–9), 1520–1528 (2009)
5. T. Hoffmann, G. Doorribos, I. Ferain, N. Collaert, P. Zimmerman, M. Goodwin, R. Rooyackers, A. Kottantharayil, Y. Yim, A. Dixit, K. De Meyer, M. Jurczak, S. Biesemans, GIDL (gate-induced drain leakage) and parasitic schottky barrier leakage elimination in aggressively scaled HfO_2/TiN FinFET devices, IEDM Tech. Dig. 725–728 (Dec 2005)
6. T. Hori, Drain-structure design for reduced band-to-band and band-todefect tunneling leakage, VLSI Technol. Symp. Tech. Dig. 69–70 (June 1990)
7. R.H. Dennard, F.H. Gaensslen, H.-N. Yu, V.L. Rideout, E. Bassous, A.R. Leblanc, Design of ion-implanted MOSFET's with very small physical dimensions. IEEE J. Solid State Circuits **9**(5), 256–268 (1974)
8. System Drivers Chapter, Updated Edition, International Technology Roadmap for Semiconductors (2010), http://www.itrs.net/Links/2010ITRS/Home2010.htm
9. Design Chapter, Updated Edition, International Technology Roadmap for Semiconductors (2010), http://www.itrs.net/Links/2010ITRS/Home2010.htm
10. Executive Summary, Updated Edition, International Technology Roadmap for Semiconductors (2010), http://www.itrs.net/Links/2010ITRS/Home2010.htm
11. Emerging Research Devices and Materials Chapters, Updated Editions, International Technology Roadmap for Semiconductors (2010), http://www.itrs.net/Links/2010ITRS/Home2010.htm

Part I
Background

Chapter 2
Background and Motivation

Abstract In this chapter, we aim to present the background and the motivation of this thesis work. We will first give an overview of the Field Programmable Gate Array architecture, which is today the most widely used reconfigurable circuit. After describing its conventional structure, we will detail current trends in architectural organization. Then, we will survey the literature to see how disruptive technologies are used to propose drastic evolutions in the field. We will in particular show how dense nanowires can be used to build logic fabrics in a crossbar organization, and also how the use of carbon electronics allows the construction of interesting logic functionalities. Finally, we will try to formalize the various approaches into a hierarchical representation and compare it to the conventional structure. This representation will help to define the objectives of this work. We mainly intend to propose a digital reconfigurable circuit based on real-life disruptive technologies. This is an important point, since even if a potential technology opens the way towards new phenomena, it is fundamental to work closely with technologists and to keep in mind its feasibility from an industrial perspective. In this context, we will continuously try, in this thesis work, to take into account the technology requirements when designing a circuit.

As introduced previously, reconfigurable logic architectures are generic and highly versatile. This makes them an excellent compromise between costs, development time and performances. Suited for a wide range of application, they offer an intrinsic regularity compatible with the most advanced technological processes.

In this state of the art chapter, we will give an overview of the reconfigurable field. Hence, we will first introduce the conventional *Field Programmable Gate Array* (FPGA) architectures. From the basic FPGA scheme, we will move to the most recent evolutions and discuss the structural issues.

Subsequently, from the emerging technologies perspective in the context of computation, we will survey the nanodevices-based architectures relevant to our field and give a global comparison between the different approaches.

P.-E. Gaillardon et al., *Disruptive Logic Architectures and Technologies*,
DOI: 10.1007/978-1-4614-3058-2_2,

Finally, we will formalize our research methodology, by describing the chosen global architectural template, as well as the position of our work regarding the existing literature.

2.1 Conventional Reconfigurable Architecture Overview

Reconfigurable architectures are today leads by the Field Programmable Gate Array. We will begin this overview with the survey of the homogeneous FPGA architectural scheme and its related sizing. Then, we will see how the homogeneous architecture has been improved to increase the structural performances. Finally, we will discuss the current limits of FPGAs.

2.1.1 The Field Programmable Gate Array Architecture

In this section, we provide an overview of the standard FPGA scheme. We will start with some generalities of the structure. These generalities will deal with a short history/overview of reconfigurable circuits and will present the basements of the FPGA architecture. Then, we will detail the logic blocks architecture. Logic blocks architectures can be based on logic gates or *Look-Up Tables* (LUTs). Interconnection structures will then be detailed, and finally, sizing of the architecture will be presented.

2.1.1.1 Generalities

Field Programmable Gate Array belongs to the family of reconfigurable logic circuits. Its structure is currently the most advanced of the family.

Historically, the reconfigurability has been based on programmable diode logic. Second generation architectures used in *Programmable Array Logic* (PAL)/ *Programmable Logic Array* (PLA) architectures [1]. The PAL approach focused on the use of a reconfigurable full interconnectivity pattern for the implementation of the signal routing between macro-logic blocks. Hence, the PAL approach is intimately defined by its large routing array. It is important to note that in such a circuit, the logic is fixed and only the routing part is programmed.

The FPGA breaks this model by using both programmable logic and programmable routing structure. The logic is distributed through the routing structure in an island-style manner. This distribution helps in handling the routing congestions with an optimum number of resources.

The PAL routing architecture is a very simple but highly inefficient crossbar structure. Every output is directly connectable to every input. Connection is made through a programmable switch. The FPGA routing architecture provides a more efficient routing where each connection typically goes through several switches. In a programmable logic device, the logic is implemented using two-level AND-OR logic with wide input for the AND gates. In an FPGA, the logic is implemented using multiple levels of lower fan-in gates, which is often much more compact than two-level implementations. An FPGA logic block could be as simple as a transistor or as complex as a *Digital Signal Processor* (DSP) block. It is typically capable of implementing many different combinational and sequential logic functions.

An FPGA structure is defined by fine-grain logic. This name comes from the granularity which is achieved by the logic blocks. Here, each logic blocks is supposed to realize a part of combinational or sequential logic operations but is not able to handle complex logic operation. The granularity is then in opposition to Massively-Parallel Processor Arrays or MultiProcessor System-On-Chip. Another particularity of the FPGA architecture is the hierarchical logic stratification. Indeed, a logic block is build from smaller logic blocks and a local reconfigurable interconnect. This interconnect is generally complete. This means that, like in a PAL, each signals (inputs and outputs) can be routed everywhere.

The routing architecture of an FPGA could be as simple as a nearest neighbor mesh [2] or as complex as the perfect shuffle used in multiprocessors [3]. More typically, an FPGA routing architecture incorporates wire segments of varying lengths which can be interconnected via electrically programmable switches. The choice of the number of wire segments incorporated affects the density achieved by an FPGA. If an inadequate number of segments is used, only a small fraction of the logic blocks can be utilized, resulting in poor FPGA density. Conversely the use of an excess number of segments that go unused also wastes area. The distribution of the lengths of the wire segments also greatly affects the density and performance achieved by an FPGA. For example, if all segments are chosen to be long, implementing local interconnections becomes too costly in area and delay. On the other hand if all segments are short, long interconnections are implemented using too many switches in series, resulting in unacceptably large delays.

The storage of the configuration is in charge of memories. These memories drive logic gates or pass-transistors in order to configure the logic or the routing. Different type of technologies could be used like *Static Random Access Memories* (SRAM), antifuse or flash. SRAM is actually the most standard technology, thanks to its CMOS technological homogeneity. In all cases, a programmable switch occupies larger area and exhibits much higher parasitic resistance and capacitance than a simple via. Additional area is also required for programming circuitry. As a result the density and performance achievable by today's FPGAs are an order of magnitude lower than that for ASIC manufactured in the same technology.

The complexity of FPGA has surpassed the point where manual design and programming are either desirable or feasible. Consequently, the utility of FPGA architecture is highly dependent on effective automated logic and layout synthesis tools to support it. A complex logic block may be under-utilized without an effective logic synthesis tool, and the overall utilization of an FPGA may be low without an effective placement and routing tool.

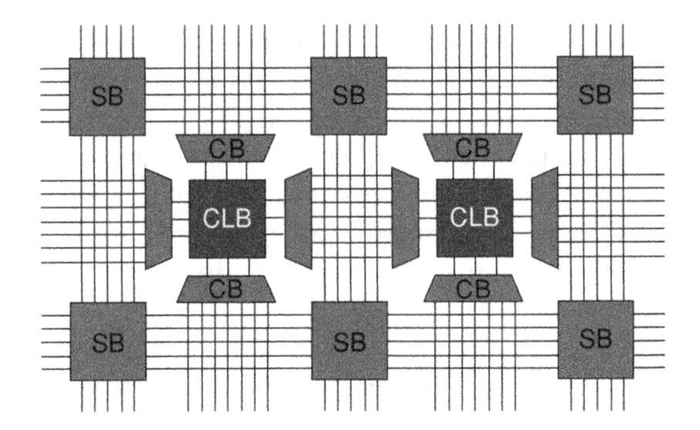

Fig. 2.1 Island-style field programmable gate array organization

Placement and routing tools mainly depend on the architecture and are specific to manufacturers [4, 5]. Some tools aims to be used as an exploration tool for research exploration of algorithms as well as architectures. As an example, the *Verilog-To-Routing* (VTR) toolflow is dedicated to this purpose [6].

2.1.1.2 General Architectural Organization

FPGAs are built of three fundamental components: logic blocks, I/O blocks and programmable routing, as shown in Fig. 2.1. In Fig. 2.1, CLB stands for *Configurable Logic Block*, which is the combinational and sequential logic block. CB and SB stand respectively for *Connection Box* and *Switch Box*. These circuits form the global routing resources. In Fig. 2.1, Input/Output (I/O) blocks have not been shown. They are found at the periphery of the circuit and are fed by the routing lines.

In an FPGA, a circuit is implemented by programming each of the logic blocks. Each block implements a small portion of the logic required by the circuit. The programmable routing is configured to make all the necessary connections between logic blocks and from logic blocks to I/O blocks, as well. Each of the I/O blocks is configured to act as either an input pad or an output pad.

2.1.1.3 Logic Block Architecture

The logic block used in an FPGA strongly influences the circuit speed and area-efficiency. Many different logic blocks have been used in FPGAs, but it is possible to consider two main families: the gate based FPGAs and the *Look-Up Table* (LUT) based FPGAs. Most current commercial FPGAs use logic blocks based on LUTs.

- Gate-based Approach

Gate-based FPGAs closely resemble basic *Application Specific Integrated Circuit's* (ASIC) cells. The finest grain logic block would be identical to a basic cell of an ASIC and would consist of A few transistors that can be interconnected

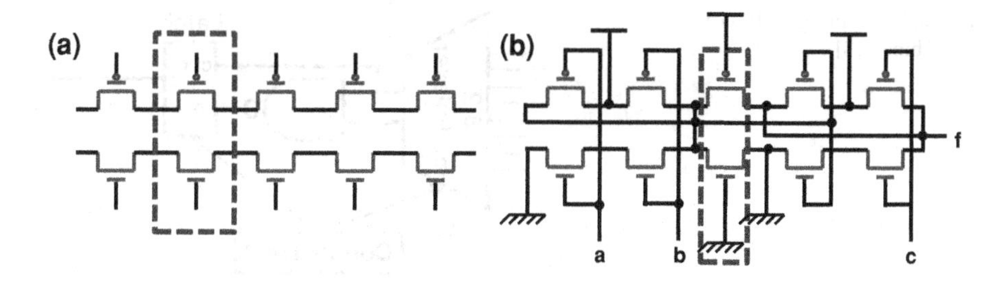

Fig. 2.2 Transistor-level programmability for FPGA [7]. **a** Transistor pair, **b** Isolation transistors

in a programmable way. The finest grain cell solution uses single transistor pairs. In [7], transistors pairs are connected together in rows. Within this pattern, the transistors are programmed to serve as isolation transistors or logic gates. Figure 2.2 illustrates how a function could be implemented. Figure 2.2a shows the transistor pair tiles and Fig. 2.2b shows a programmed function $f = a{\cdot}b + !c$. The function is programmed by ensuring the correct connections between the different transistors. In Fig. 2.2b, the dashed lines show the transistors that are turned off for isolation. The function is done by the two-input NAND gates formed on the left and right sides.

Instead of using programmability at the lowest transistor level, a two-input NAND gate has been used to realize the combinational block, as depicted in Fig. 2.3 [8]. The expected logic function is formed in the usual way by connecting the NAND gates together.

The main advantage of using fine grain logic blocks is that these blocks are fully utilized. This is because the logic synthesis techniques for such blocks are very similar to those for conventional mask-programmed gate arrays and standard cells. Then, it is easier to use small logic gates efficiently. The main disadvantage of these blocks is that they require a large number of wire segments and programmable switches. Such routing resources are costly in terms of delay and area. As a result, FPGAs employing fine-grain blocks are in general slower and achieve lower densities than those employing coarse grain blocks.

In [9], more complex logic blocks are proposed. Figure 2.4a shows that these blocks are based on the ability of a multiplexer to implement different logic functions by connecting each of its inputs to a constant or to a signal. Reference [10] presents a similar solution. Each input of the multiplexer and not just the select input is driven by an AND gate, as illustrated in Fig. 2.4b.

The alternating inputs to the AND gates are inverted. This allows input signals to be passed in true or complement form, thus eliminating the need to use extra logic blocks to perform simple inversions. Multiplexer-based logic blocks have the advantage of providing a large degree of functionality for a relatively small number of transistors. This is, however, achieved at the expense of a large number of inputs (8 in the case of Actel and 20 in the case of QuickLogic), which when utilized place high demands on the routing resources.

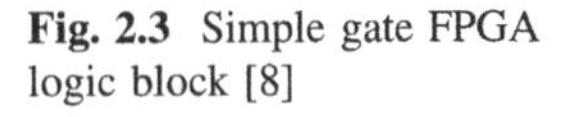
Fig. 2.3 Simple gate FPGA logic block [8]

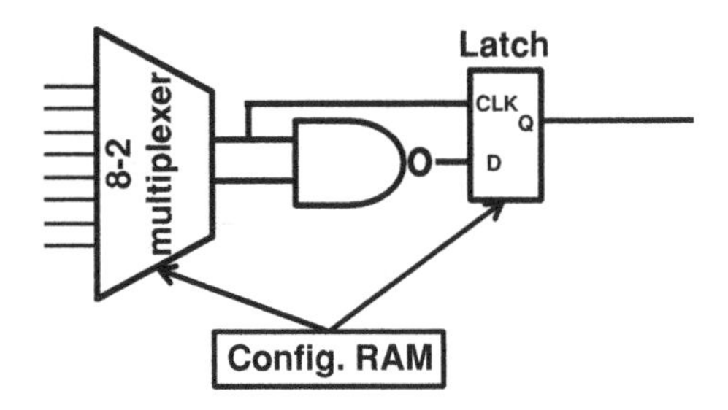

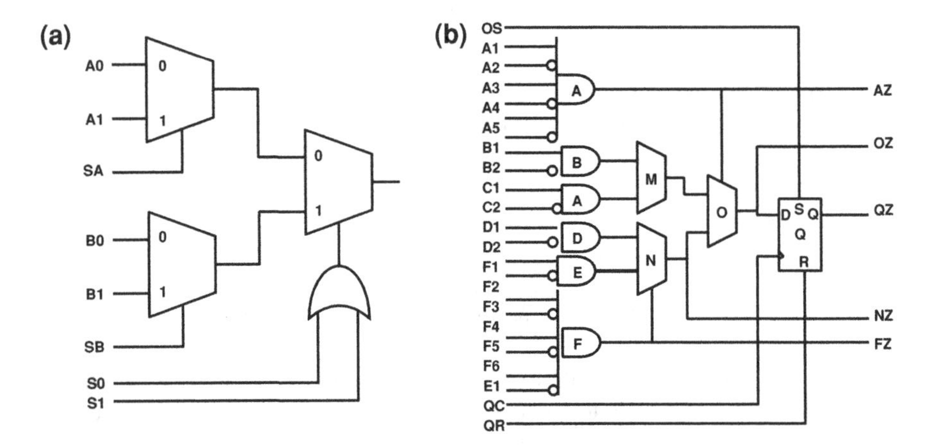

Fig. 2.4 Complex gate FPGA logic blocks. **a** Actel Act-1 LB [9]. **b** Quicklogic pASIC1 LB [10]

- Look-Up Table-based Approach

The other and most used approach is based on LUT. A LUT works thanks to memories driving the data inputs of the multiplexer. The truth table for a K-input logic function is stored in a $2^K \times 1$ SRAM. The address lines of the SRAM function as inputs and the output of the SRAM provides the value of the logic function. For example, we consider the logic function $f = a{\cdot}b + !c$. If this logic function is implemented using a three-input LUT, then the SRAM would have a 1 stored at address 000, a 0 at 001 and so on, as specified by the truth table. The advantage of look-up tables is that they exhibit high functionality. A K-input LUT can implement any function of K-inputs, i.e. 2^{2^K} functions. The disadvantage is that they are unacceptably large for more than about five inputs, since the number of memory cells needed for a K-input lookup table is 2^K. While the number of functions that can be implemented increases very fast, these additional functions are not commonly used in logic designs and are difficultly handled by the logic synthesis tools. Hence, it is often the case that a large LUT will be largely under-utilized.

Most modern FPGAs are composed not of a single LUT, but of groups of LUTs and registers with some local interconnect between them. A generic view for the LUT based logic block is shown in Fig. 2.5.

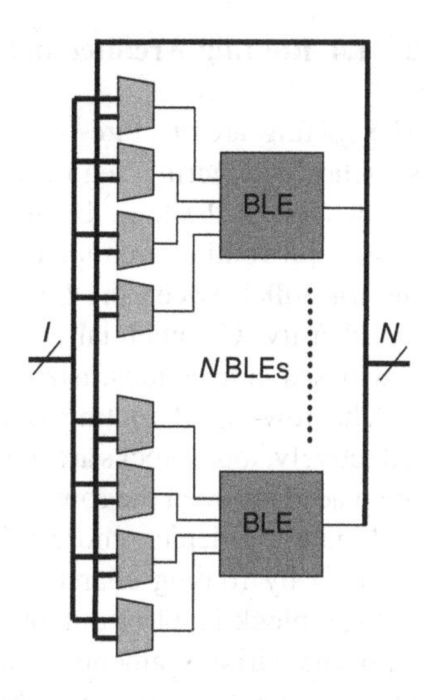

Fig. 2.5 Configurable logic block architecture [11]

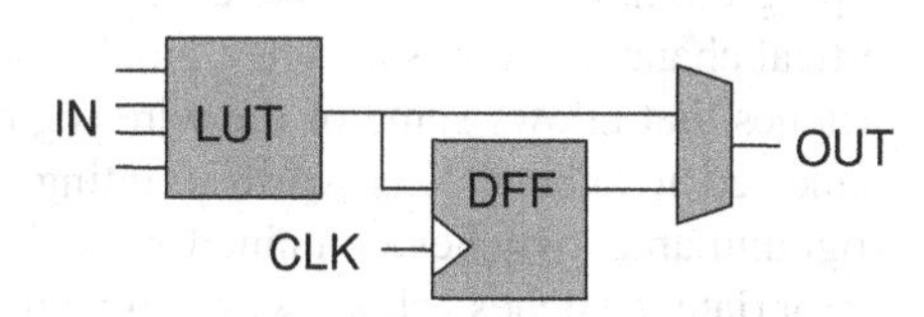

Fig. 2.6 Basic logic element architecture [11]

This CLB has a two-level hierarchy: the overall block is a collection of *Basic Logic Elements* (BLEs) [11]. As shown in Fig. 2.6, the BLE consists of a LUT and a register. Its output can be either the registered or unregistered version of the LUT output.

This is how many commercial FPGAs combine a LUT and a register to create a structure capable of implementing either combinational or sequential logic. The complete logic block contains several BLEs and local routing to interconnect them. Such a generic logic cluster scheme is described by three parameters [11]: the number of inputs to a LUT (K), the number of BLEs in a cluster (N), the number of inputs to the cluster for use as inputs by the LUTs (I). It is worth noticing that not all $K{\cdot}N$ LUT inputs are accessible from outside the logic cluster. Instead, only I external inputs are provided to the logic cluster. Multiplexers allow arbitrary connections of these cluster inputs to the BLE inputs. They also allow the connection of all the N outputs to each of the BLE inputs. All the N outputs of the logic cluster can be connected to the FPGA routing for use by other logic clusters. It is remarkable that each of the BLE inputs can be connected to any of the cluster inputs or any of the BLE outputs. Logic clusters are therefore internally fully connected. This is a useful feature, as it simplifies *Computer Aided Design* (CAD) tools considerably.

The presented structure is a very generic representation of LUT-based FPGAs. In fact, logic blocks of FPGAs are more complex, as detailed in the next sub-chapter.

2.1.1.4 Routing Architecture

The routing architecture of an FPGA is the manner in which the programmable switches and wiring segments are positioned to allow the programmable interconnection of the logic blocks.

Several routing architectures exist and come most of the time from results on the tradeoff between the flexibility of the routing architecture, circuit routability and density. Commercial routing approaches can be classified into three groups: row-based connections, island-style connections, and hierarchical scheme.

The row-based routing scheme is close to ASIC standard cells routing [9]. Effectively, logic blocks are organized in rows and a large number of horizontal wires are placed between the rows. Less vertical wires are used to connect rows to others.

Figure 2.7 depicts more precisely an island-style FPGA. Logic blocs are surrounded by routing channels of pre-fabricated wire segments on all four sides. A logic block input or output, which is called a pin, can be connected to some or all of the wiring segments in the channel adjacent to it via a connection block [12] of programmable switches. At every intersection of a horizontal channel and a vertical channel, there is a switch block [12]. This is simply a set of programmable switches that allows some of the wire segments incident to the switch block to be connected to others. It is worth pointing out that in Fig. 2.7, only a few of the programmable switches contained by switchboxes are shown. By turning on the appropriate switches, short wire segments can be connected together to form longer connections. In the figure, some wire segments continue unbroken through a switchbox. These longer wires span multiple logic blocks, and are a crucial feature in commercial FPGAs.

The number of wires contained in a channel is denoted by W. The number of wires in each channel to which a logic block pin can connect is called the connection block flexibility, or F_c. The number of wires to which each incoming wire can connect in a switch block is called the switch block flexibility, or F_s.

Inspired from generic programmable logic devices routing schemes, the hierarchical routing scheme try to use some form of locality to obtain better density and performance. This hierarchy could be realized with several different interconnect schemes, like PAL or island-style. This is the most currently used interconnection in FPGAs. For example, considering the structure presented in Figs. 2.1 and 2.5, we could remark that a full-interconnectivity scheme is used for CLB, while island-style organization is used for the global FPGA.

2.1.1.5 Architectural Parameterization and Optimum

The programmable switches introduced for routing purpose impact the performances. Thus, the FPGA architecture is the results of a trade-off between high versatility and cost/performance.

The adverse effects of the large size and relatively high parasitic of programmable switches can be reduced by careful architectural choices. By choosing the

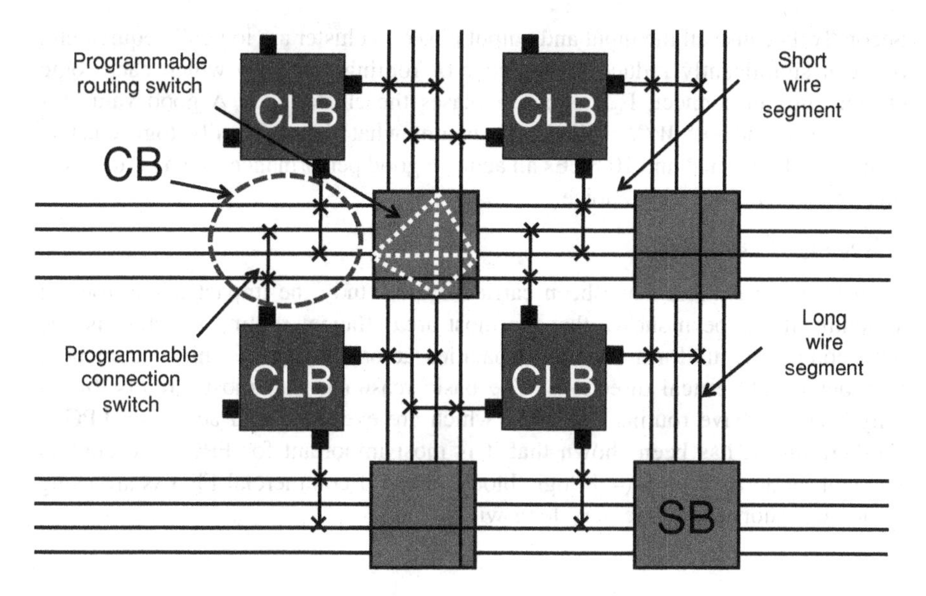

Fig. 2.7 Details on the routing organization in an island-style FPGA [11]

appropriate granularity and functionality of the logic block, and by designing the routing architecture to achieve a high degree of routability while minimizing the number of switches, both density and performance can be optimized. The best architectural choices, however, are highly dependent on the programming technology used as well as on the type of designs implemented, so that no unique architecture is likely to be best suited for all programming technologies and for all designs.

A complete study of architectural parameterization has been conducted in [11, 13, 14]. These studies assume SRAM based homogeneous FPGAs. The principal results are summarized in the following.

- Combinational Granularity Impact

As the granularity of a logic block increases, the number of blocks needed to implement a design should decrease. On the other hand a more functional (larger granularity) logic block requires more circuitry to implement it, and therefore occupies more area. This tradeoff suggests the existence of an "optimal" logic block granularity for which the FPGA area devoted to logic implementation is minimized. It has been shown in [13] that the most suited LUT size is reached for K equal to 4.

- Logic Block Sizing

Several interesting sizing results can be taken from [11]. Firstly, the number of distinct inputs required by a logic cluster grows fairly slowly with cluster size, N. A cluster of size N requires approximately $2N + 2$ distinct inputs (for $N \leq 20$).

Secondly, because all the input and output pins of a cluster are logically equivalent, one can significantly reduce the number of routing tracks to which each logic cluster pin can connect, F_c, as one increases the cluster size. A good value for $F_{c,output}$, is found with *W/N*, while $F_{c,input}$ is somewhat higher. Thirdly, logic clusters containing between 4 and 10 BLEs all achieve good performance, so any clusters in this range is a reasonable choice.

- Routing Organization

In [11], simulations have been carried out to study the impact of the routing structure. It has been shown that the most area-efficient routing structure is one with completely uniform channel capacities across the entire chip and in both horizontal and vertical directions. The basic reason is that most circuits "naturally" tend to have routing demands, which are evenly spread across an FPGA. Furthermore, it has been shown that it is most important for FPGAs to contain wires of moderate length (4–8 logic blocks) even if commercial FPGAs are using some very short and some very long wires.

2.1.2 Market Trends

In the previous sub-chapter, we focused on the FPGA architecture and especially the generic homogeneous island-style architecture. While the island-scheme is used in most FPGAs, their structures are relatively complex. Indeed, the generic architecture is useful for understanding and research purpose, but it suffers from several performance issues in some application classes. Thus, several improvements have been proposed for modern commercial FPGA. The circuits are enhanced in many ways.

2.1.2.1 Increase of the CLBs Complexity

In order to increase the logic block functionality, the structure has been customized and new features have been added. The logic block of a modern Xilinx Virtex-6 FPGA is shown in Fig. 2.8.

The block is still organized around LUTs and FFs. These elements give the combinational and sequential ability. The first novelty is the fractional behavior of the LUT. A large LUT is useful to realize a large and complex combinational function. However, this situation is not frequent. Instead of wasting a large amount of logic resources when the LUT implements a small function, it is possible to split it into two independent smaller LUTs. It is then possible to optimize the logic fabric to the application, during the synthesis and packing operations.

Within the CLB, extra circuitries have been added to address specific functionalities. We can point out in particular notice the carry lines, the SRAM decoders and the shift-register.

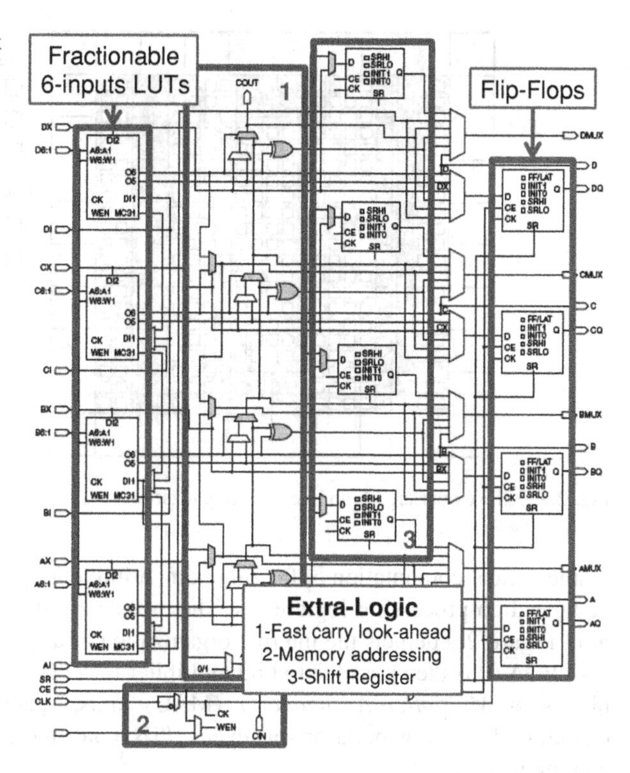

Fig. 2.8 Virtex-6 logic block [15]

The carry lines extend through several LBs, in order to implement logic multipliers efficiently. Effectively, the specific track avoid to implement the carry through the global interconnect. Since global interconnect must be routed, delays for carry will be unknown and in any case larger than with the specific track. This allows significant increase in the performance of such a block without costing too many resources.

The use of FPGAs keeps increasing to implement complex SoCs. Thus, the requirement in terms of embedded memories is growing. LBs embed a large amount of SRAM memories for the LUT configuration. It is then of high interest to give the possibility to use them directly as standalone memories. In this configuration, the LUTs multiplexers are used as the address decoders, and a specific extra circuitry deals with the read management of the read/write control signals.

Finally, shift register behavior is also achievable, thanks to a dedicated module. Indeed, SRAMs can be cascaded and the writing is driven by a shifting register clock signal. This is obviously precious for specific applications.

2.1.2.2 Transition from Homogeneity to Heterogeneity

Associated with the complexity and diversity trends of the CLB architectures, the FPGA scheme tends to add more dedicated blocks and to heterogenize its structure. Indeed, the homogeneous FPGA is known to be slow and area costly for

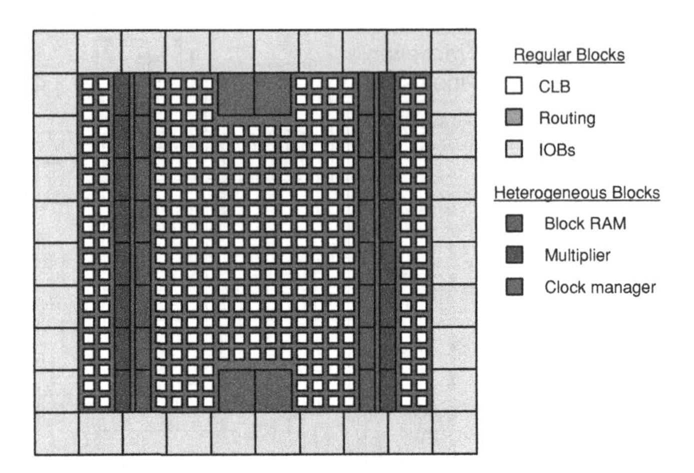

Fig. 2.9 Xilinx Spartan-3A architecture organization [16]

mathematical computation or floating point arithmetic. Hence, vendors have added dedicated co-processing logic units. These units are distributed through the logic grid. Figure 2.9 depicts the internal organization of a commercial Xilinx Spartant-3A FPGA. It is then possible to find multiplier for *Digital Signal Processing* (DSP) blocks or *Multiplier–Accumulator* (MAC) units, specific blocks for complex clocking domain generation or even USB controller, Ethernet controller for communications.

Furthermore, requirements on memories are wide in current applications. FPGAs then integrate dedicated standalone RAM blocks directly reachable by the routing part. This allows the creation of complex SoC and the efficient instantiation of soft microprocessors on the structure.

2.1.2.3 Non-Volatility Features

In high production FPGAs, the configuration is stored by SRAM memories. These memories are found distributed through the entire circuits to store the information as close as possible to the data path logic. SRAMs circuits use the same integration process than the other logic parts. Thus, they are the simplest solution to implement distributed configuration logic. Nevertheless, they suffer from their large size and, in particular from their volatile behavior.

Due to the volatility, the configuration must be reloaded into the FPGA circuits at each power-up. This obviously leads to a large loss of efficiency in terms of delay and power consumption. Furthermore, a non-volatile storage is still required outside of the chip. In current circuits, standalone flash memories are used to store the configuration bitstream, and specific programming circuits have in charge the programming sequence. This is highly area and power consuming. Then, it appears suited to use on-chip non-volatile memories directly.

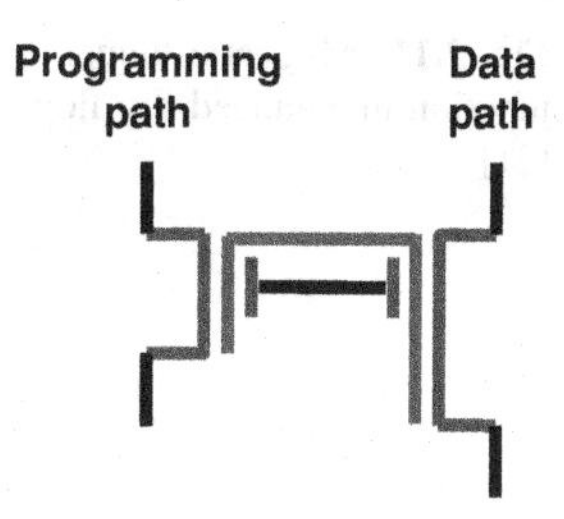

Fig. 2.10 Schematic of a merged programming-data path floating gate transistor [20]

• Flash-based Approach

The current dominant non-volatile memory technology is the flash technology. Flash requires several technological steps in addition to CMOS. To take into account the complexity of flash/CMOS co-integration, the simpler solution is to co-integrate in the same package or directly on the same die the SRAM FPGA and its configuration flash memory. This solution is used in [16]. Such integration decreases the extra-chip requirements. Nevertheless, the most important hurdles still exist. Information is still duplicated in the flash and in the FPGA's SRAMs. This means that power consumption remains the same with a large wasted power in the distributed RAMs and a long power up time.

Thus, solutions have been envisaged to distribute the non-volatile memories through the logic grid. Flash based solution is proposed in [17–20], while emerging solutions are shown in [21].

In [17], co-integration of flash and MOS transistors yield in a compact configuration memory nodes. The structure uses a non-volatile pull-up (pull-down respectively) network and resistive pull-down (pull-up respectively) network. This voltage divider arrangement allows the storage of the configuration and the drive of logic gates. This enables the fabrication of a non-volatile base LUT, as introduced in [18]. Complex co-integration permits to create compact non-volatile switches for FPGAs. In fact, flash transistor could be seen as a programmable switch. The programming of a flash transistor needs specific voltages and circuitries. Traditionally, these circuits would be introduced into the data path. In [19, 20], a simple circuit composed of two flash transistors is presented (Fig. 2.10). The particularity is that the floating gate is shared in order to connect a dedicated programming transistor with the data path one. This is of high interest for logic compactness, and distinct separation between programming and logic path.

• Magnetic Memory-based Approach

In [21], magnetic RAMs are used instead of flash. Indeed, emerging resistive memories are expected to reduce the production costs of the circuits. Resistive memories could be integrated after the back-end process. This means that the resistive devices are realized only after the costly CMOS process. This is obviously interesting for cost reduction and process simplification. Figure 2.11 depicts the circuit that is used to store the information. The circuit is unbalanced flip-flop. Two magnetic tunnel junction memories store complementary data that are used to

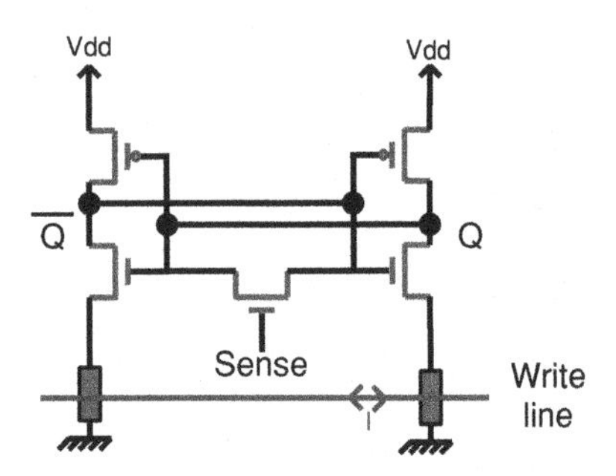

Fig. 2.11 Magnetic tunnel junction unbalanced flip flop [21]

start the flip-flop in a good configuration at power up. The programming of the magnetic memories is in charge of special writing lines. The magnetic memories are placed above the IC as explained in [21]. Nevertheless, we should remark that this solution is area consuming due to the presence of front-end transistors in addition to the size of the back-end memories.

2.1.3 Limitations of FPGAs

Figure 2.12 presents the area/delay/power breakdown of the various components of a baseline Xilinx Virtex island style FPGA.

The impact of area of each FPGA part has been objectively studied in [22]. This evaluation has been realized with a precise methodology using a stick diagram of the implemented circuits, instead of using a simpler minimum-width equivalent size transistor count. It is worth noticing that the configuration memories occupy roughly half of the area in both the logic blocks and the routing resources. The logic blocks occupy only 22% of the whole area including its own configuration memory. Only 14% is then used for actual computation.

In addition to consuming most of the die area, programmable routing also contributes significantly to the total path delay in FPGAs. In [23, 24], interconnect delays are estimated and found to account for roughly 80% of the total path delay. Programmable routing also contributes to the high power consumption of FPGAs. This problem has recently become a significant impediment to the FPGA adoption in many applications. The power consumption measurements of some commercial FPGAs have shown that programmable routing contributes more than 60% of the total dynamic power consumption [25–27]. As a result of these performance degradations, FPGA performance is significantly worse in terms of logic density, delay, and power than cell-based implementations. Finally, it is commonly admitted that FPGAs are more than ten times less efficient in logic density, three times larger in delay, and 3 times higher in total power consumption than cell-based implementations [28].

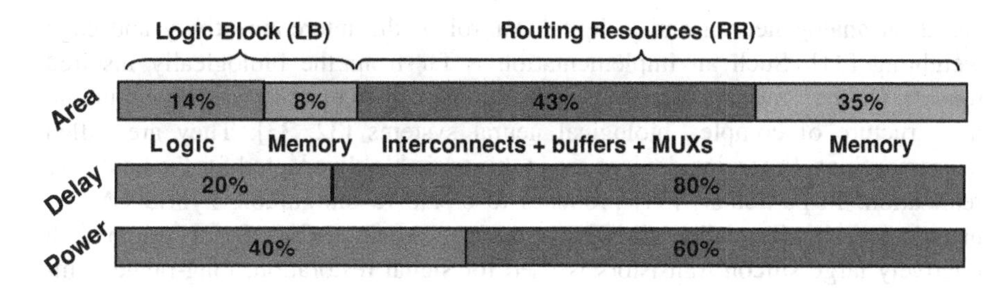

Fig. 2.12 Field programmable gate arrays area/delay/power repartition per block [22]

2.2 Emerging Reconfigurable Architectures Overview

While we expect that reconfigurable architecture represent the future of computation architectures, we will assess the ways in which emerging devices could improve or even break the current FPGA model. This represents a difficult challenge because in many cases, circuit-level models and/or architecture-level models for these devices and their interconnect systems either do not exist or they are very primitive. Moreover, the applications under considerations for these new devices can take many forms; i.e.

- as a drop-in replacement for CMOS,
- as additional devices that complement and coexist with CMOS devices, or
- as devices whose unusual properties can provide unique functionality for selected information processing applications.

The ITRS—Emerging Research Devices [29] and especially the *Emerging Research Architectures* section aims to identify possible applications for emerging logic and memory devices. Two main families of architectures are defined: the morphic architectures and the heterogeneous architectures. We will survey the principal architectures that have been devised for computing at the nanoscale.

2.2.1 Morphic Approach

As indicated by the etymology, the morphic approach aims to develop architectures that are inspired from other systems. They are often inspired from biology as far as biological systems are highly efficient due to very limited power consumption and high system reliability. In this class of application, neural networks immediately spring to mind. An *Artificial Neural Network* (ANN) is a model inspired by the human brain [30]. Its goal is to reproduce some properties found in the biological organ. Globally, ANNs are used in the following applications which benefit from its properties related to memories and/or learning: sorting, associative memories, compression… The conventional implementation of neural networks is based on analog neuron and a digital control of the interconnectivity and edge weighting [31]. Such an implementation is far from the biologically inspired models. On the contrary, some implementations copy to emulate the behaviour and the structure of complex biological neural systems [32, 33]. They are called neuromorphics. Emerging devices are in this topic highly adapted for the hardware realization. In particular, [34] proposes to create reconfigurable hybrid CMOS/nanodevice circuits, called CMOL. In such a circuit, a CMOS subsystem with relatively large silicon transistors is used for signal restoration, long-range communications, input/output functions, and testing/bootstrapping. An add-on nanowire crossbar with simple two terminal nanodevices at each cross-point provides most of information storage and short-range communications.

The emergence of new devices and integration paradigms will certainly lead to several improvements of neuromorphic circuits. In this book, we are focusing on standard computation paradigms. Thus, we will consider that architectures suited for morphic applications are outside of our scope.

2.2.2 Heterogeneous Approach

Emerging technologies are expected to supersede CMOS in terms of functionality and performance metrics. Nevertheless, the transition to a new disruptive technology will not occur abruptly. In the near future, it seems reasonable to consider that standard CMOS circuits will be improved by new technologies. Two different means of improvement can be pointed out: the improvements coming from the increase of the device functionality (i.e. the use of devices with more functionality in the same area) and an improvement coming from the increase of integration density (i.e. more devices in the same area).

2.2.2.1 Regular Architectures

Mono-dimensional devices improve the performance of transistors channel. Nevertheless, the ultra-scaled dimensions represent a real challenge for the integration of complex circuits. In particular, photolithography unreliability requires the use of high regularity. Regularity can be envisaged at the transistor level. The use of micro-regularity is of high interest to reduce the size of the circuits drastically. Furthermore, regularity is compatible with bottom-up fabrication techniques. These techniques open the way towards complex arrangements at the nano-scale and lead to the emergence of crossbar circuits. A crossbar is defined by the regular arrangement of devices in the array. This leads to the most integrated structures achievable by the technology.

The first crossbars realized from emerging devices were proposed in [35–37]. The basic function that those circuits are implementing is information storage. Subsequently, the use of dense crossbars computational units has been conceptually

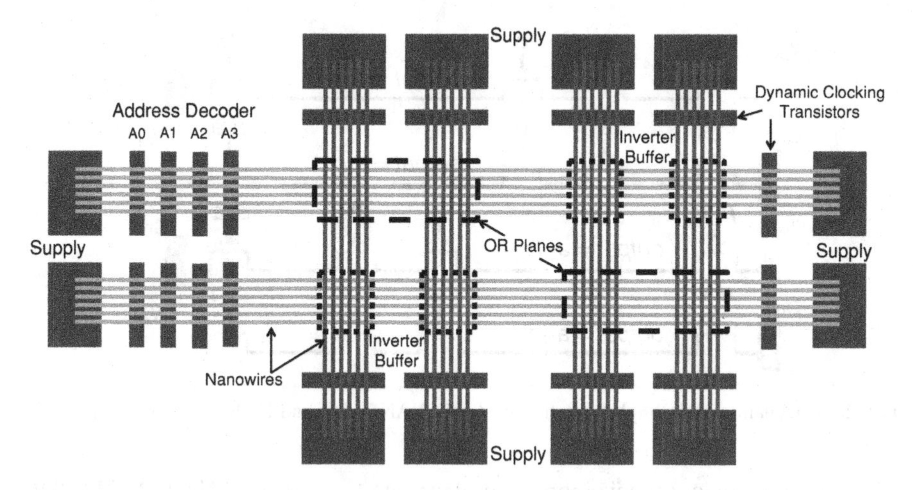

Fig. 2.13 NanoPLA architecture [41]

proposed in [38–40]. More precisely, such computation fabrics are based on semiconducting *Silicon NanoWires* (SiNWs) organized in a crossbar fashion. Active devices are formed at the cross-points. In [41], a Programmable Logic Array is proposed. The cross-points are realized by molecular switches. The switches can be programmed in order to perform either signal routing or wired-OR logic function. The structure is arranged in several sub-crossbars and presented in Fig. 2.13.

The input of the structure is a decoder interface between the micro/nano worlds. The decoder is used to address every nanowire independently of the others. The decoder design assumes that the nanowires are differentiated by a given doping profile [42]. The output of the crossbar is routed to a second crossbar. The signals can be inverted by gating the nanowires carrying the signals. A cascade of these two planes is equivalent to a NOR plane. It is worth noting that, due to diode logic, a logic restoration stage is required. Indeed, a more than unity gain is required to ensure a good cascade. At the end, the structure is duplicated many times and several stages are connected to each other in order to perform complex logic functions.

Instead of using diode logic, [43] uses FETs realized at the cross-points. While the previous approach implements a reconfigurable PLA circuit, this technology based on FETs address specific application-driven designs. Within this organization, a *Nanoscale Application Specific Integrated Circuit* (NASIC) is introduced in [44]. A NASIC tile consists of basic circuits such as adders, multiplexers and flip-flops. Circuits are realized using a dynamic logic style. Two clock transistors are placed between the power lines and a stack of transistors, which realize the logic function. It is possible to implement a standard AND/OR functions and their inverted counterparts. The implementation of the different logic functions is depicted in Fig. 2.14.

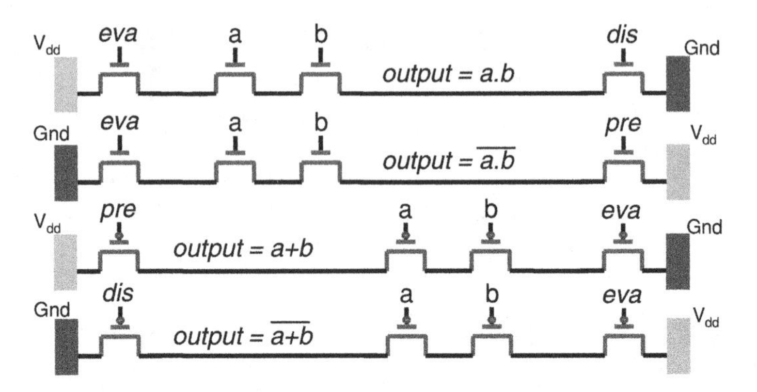

Fig. 2.14 Dynamic logic implementation of AND, NAND, OR and NOR functions [44]

From this logic organization, two-plane are cascaded (AND/OR, NAND/NAND…). Figure 2.15 depicts the complete NASIC tile. A first set of AND logic functions is realized in the horizontal direction. Nano-scale wires are connected to micro-scale power lines and to the other blocks that are surrounding the crossbar core. The second horizontal AND functions are driving transistors into the vertical orientation. The vertical line set implements OR functions. As a complete illustration, the circuit, presented in Fig. 2.15, implements a 1-bit full adder.

The doping of nano-grid strips, the size of the NASIC tiles, the use of certain nano-scale (i.e. sub-lithographic) wires as interconnect between tiles and the micro-level interconnects are chosen in an application/architecture-domain specific manner. These aspects determine a *NASIC* fabric and are the key differentiators between PLA type of nanoscale designs [41] and NASICs.

The most sensitive issue in these crossbar proposals remains the fabrication assessments. In fact, realization of 1D structures and their alignment over long distance with a good aspect ratio is extremely difficult to perform.

Several works have been published in order to assess the technological credibility of the structure [45]. Recently, a simple programmable crossbar-based processor has been fabricated. In [46], the authors have demonstrated a programmable and scalable architecture based on a unit logic tile consisting of two interconnected, programmable, non-volatile nanowire transistor arrays. The transistor structure is depicted in Fig. 2.16. S, D and G correspond to source, drain and gate, respectively. On left, the hole concentration in a *p*-type Ge/Si *NanoWire Field Effect Transistor* (NWFET) for two charge-trapping states is presented. This charge accumulation is in charge of the hysteresis behavior of the conductance.

Hence, each NWFET node in an array can be programmed to act as an active or an inactive transistor state. This is done by charge trapping into the floating layer. By mapping different active-node patterns into the array, combinational and sequential logic functions including full adder, full substractor, multiplexer, demultiplexer and D-latch can be realized with the same programmable tile. Cascading this unit logic tile into linear or tree-like interconnected arrays is

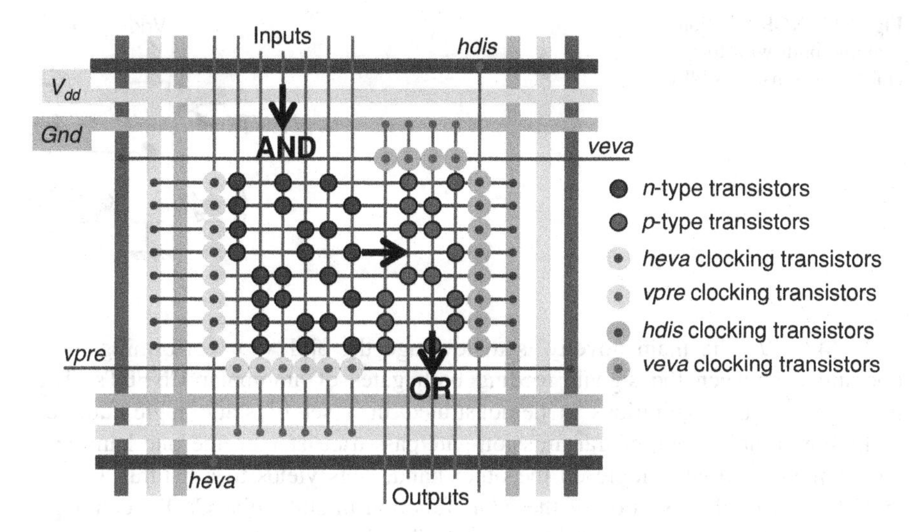

Fig. 2.15 NASIC structure (implementing a 1-bit adder) [44]

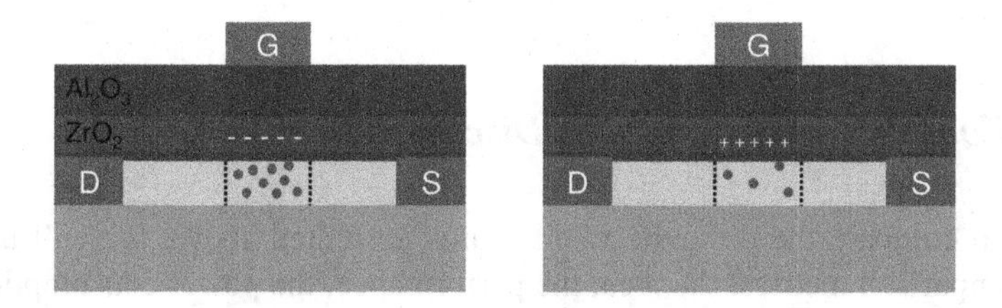

Fig. 2.16 Structure of a programmable NWFET [46]

possible given the demonstrated gain and matched input–output voltage levels of NWFET devices. This provides a promising bottom-up strategy for developing increasingly complex nanoprocessors with heterogeneous building blocks.

2.2.2.2 Enhanced Reconfigurable Architectures

In the previous part, emerging devices have been used to realize crossbar circuits. In fact, 1D structures have been mainly envisaged to create dense interconnection networks or dense substrates to build active devices at the nanoscale. Nevertheless, some 1D materials could be used efficiently to obtain new functionalities at the device level. For example, *Carbon NanoTubes Field Effect Transistor* (CNFET) exhibits an ambipolarity property [47]. This means that the same device could be controlled between *n*- or *p*-type, only thanks to the voltage applied to back-gate electrode. This property of CNTs is an opportunity that does not exist in CMOS technology. In [48–50], the benefit on logic circuit design is assessed.

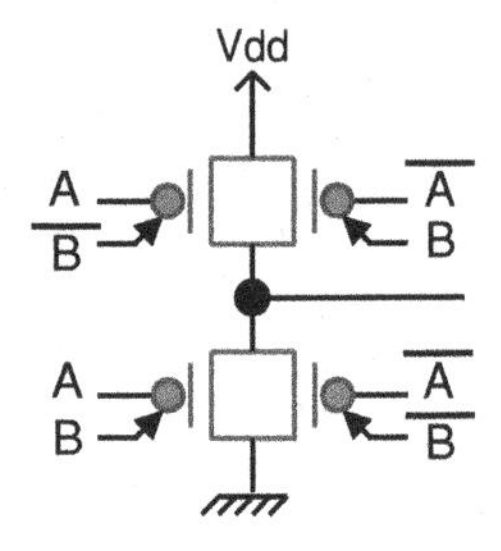

Fig. 2.17 XOR function example built with four ambipolar transistors [49]

In [48, 49], the main novelty is to leverage the ability of performing logic operations between the signals feeding both gates of ambipolar CNFETs. The design with such operations is demonstrated in a set of static logic families including combinations of transmission-gate/pass-transistor on the one hand and complementary/pseudo logic on the other hand. This yields to a natural, simple and efficient implementation of the XOR function in ambipolar CNT technology with almost no cost, as shown in Fig. 2.17. The basics of the proposition are to configure the polarity of the input signal. The polarity choice is obviously used by the configuration voltage and the transistor type.

2.2.3 Global Comparisons and Discussions

In order to compare the different architectures presented above, we will use five global comparison metrics: the area, the performance, the power consumption, the technological maturity and the fault tolerance ability. Table 1 presents the global results.

It is worth noting that a solution based on sublithographic nanowires seems to increase drastically the area of the final circuits. This seems obvious considering that the integration density of the elementary devices is enhanced. Nevertheless, these solutions are quite worse in terms of other metrics. In fact, NW-based solutions require the use of complex clocking signals to restore the logic levels. This means that a large part of performance and power are waster only for the periphery. In this context of intense power consumption, it is remarkable that carbon electronics is promising due to the good intrinsic properties of the devices. Thanks to carbon technology, it is possible to build architectural solutions that are globally efficient in all the proposed metrics, even in terms of fault tolerance and reliability. Carbon electronics have been widely studied, to enhance the reliability of the fabrication processes, and to provide complete methodologies for robust designs [51, 52]. Finally, we should consider that morphic approach provides naturally a high robustness. Effectively, these approaches are implementing neural networks. This computation scheme is well known for its versatility and performance regarding faulty tolerance.

Table 1 State-of-art architecture global metrics comparisons

	Reference publications	Area	Performances	Power consumption	Maturity	Fault tolerance
CMOL	[34]	+	+	+	– – –	+ + +
NanoFabric	[38, 39]	+ +	–	–	–	–
NanoPLA	[41]	+ +	–	–	–	–
NASIC	[43]	+ +	+	–	–	–
DG-CNFET	[49]	+	+	+ +	–	+

2.3 Architectural Formalization and Template

In this book, a fast and global evaluation methodology is presented and used to evaluate technologies from an architectural perspective. The proposed method requires the definition of a generic architectural template. From the existing state-of-the-art, it is possible to extract some global comments, regarding the design strategy and methodology.

2.3.1 Global Statement

Several works have been conducted on the fabrication technology process, reliability, circuit design, architectural definition and associated tools. Design with emerging technologies suffers from several unknown parameters. It is challenging to evaluate the expected performance of a whole circuit. Only a few works actually merge more than two aspects. Nevertheless, it is desirable to evaluate the potentiality of the technology within a complete architecture. The term complete covers the questions regarding the technological credibility of assumptions, the design of the computation part itself, the design of all the peripheral circuitries and the associated methodologies. This is required addressing all the previously defined aspects. As an illustration, the design of a compact logic gate does not make sense if the gate is not robust enough or requires significant additional circuitry.

Along the same lines, it is worth highlighting that the work on global architectures is often completely uncorrelated from the technological assumptions. This is standard architecture design with mature CMOS process. In the case of an advanced process, it is hard to handle the design of architecture without credible technological assumptions. As an example, while the use of bottom-up aligned SiNW is credible, it is difficult to imagine the wires aligned over several micrometers at only a pitch of some nanometers. This will leads to an angle deviation less than 10^{-5} degrees and is hard to handle today, as well in the near future. This makes that the architecture study must not be done under unrealistic assumptions, to ensure its credibility.

Finally, while it is recognized that variability and technological issues will increase drastically in the future, only a few works actually consider the questions of reliability. Conventional robustness techniques could be used, such as defect avoidance [53], error correction coding [54], or redundancy. Generally, unreliability in systems is overcome by duplicating the unreliable resource. Such spare circuits will then be used as a replacement part if the primary circuits are defective. However, a robust design approach is preferred to increase the reliability of a circuit early in the design [51, 52]. This allows to directly correct the misperformance at the level where it occurs, instead of a much higher level. Indeed, the correction will have a much larger and detrimental impact on the whole circuit if it is performed far from the origin of the issue. Even if the question of reliability must be tackled in future systems, we consider this specific point as out of the scope of this book.

To avoid this lack in methodology, it appears interesting to address the question of the architecture globally. Nevertheless, it is of course, not possible to handle it directly, due to the complexity of the design. Indeed, the problem concerns the handling of a complex mix between gates that leads to computational units, memories and routing with so many unknown parameters. In this book, we will use a generic template for reconfigurable architecture in order to organize the contributions into a hierarchy. The template will be presented in the following. Each layer will then be studied and optimized independently. A correct hierarchy organization ensures that the levels will not impact the others as long as interface requirements remain the same. The optimization of specific parts will be examined in the following chapters. Finally, even if a layer is improved in terms of area or functionality, it is important to assess the impact on the architectural level. Thus, it will be necessary to define a benchmarking tool which is compatible with our architectural template, and which is able to implement much different architecture for each layer. This will allow exploring the architectural design space and permitting fast track optimization for the designs. This will be presented in the following.

2.3.2 Generic Architectural Template

In this book, we are focusing on reconfigurable architectures and the enhancement that could come from emerging technologies.

Reconfigurable architectures such as FPGAs are highly versatile and adaptive in terms of target application. They use regular architecture, with several blocks that are replicated through the circuit. We have previously seen that the FPGA structure is organized hierarchically, which seems a sound approach to manage the large amount of replicated blocks. In this context, it is difficult to see where emerging technologies will be useful to improve the performance of the structure. Furthermore, the FPGA structure has been designed for CMOS-based LUT logic. The structure is optimal for this application, but we could expect that the structure will not be the same in the case of emerging technology. Thus, it is of great interest to propose an architectural

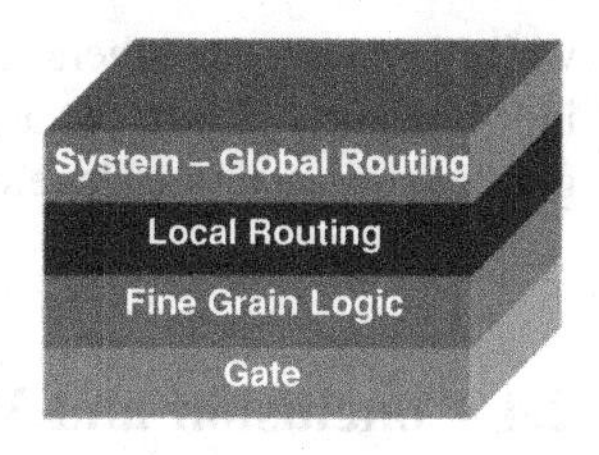

Fig. 2.18 Layered organization of the generic template architecture

template based on a hierarchical organization of configurable and routing blocks, and to generalize each hierarchical level to its global behaviour as well.

The generic template is shown in Fig. 2.18. The template is organized into four levels of hierarchy: the *Gate* level, the *Fine Grain Logic* level, the *Local Routing* and the *System-Global Routing* level.

The *Gate* level corresponds to the elementary gates that are used to perform the computation. By computation is meant the combinational and the sequential operations. Computation in FPGAs is generally realized by LUTs, but it also could be, as we have seen in the literature review, custom logic from the simple transistor to the bigger logic function.

The *Fine Grain Logic* corresponds to the smaller autonomous block that could perform both combinational and sequential logic. We consider also at this level all the configuration memories that are used to program the structure. These memories give the fully programmable functionality of the block. In the FPGA scheme, this level corresponds to the BLE, i.e. a LUT which is used to perform a fine grain operation. The LUT output is routed to a latch but also directly to an output multiplexer, which is in charge of the path selection between the latches and the unlatched version of the signal.

The *Local Routing* level aims to provide an arrangement of Fine Grain Logic and to provide it with a local connectivity pattern. Generally, this connectivity patter is full and aims to provide a large kind of programmable paths at the lower level. This allows a huge simplification of the packing tools. In the FPGA scheme, this level corresponds to the CLB. Indeed, an assembly of BLEs are fully interconnected by a multiplexer based interconnect. Thus, it is possible to realize with such a block a coarser function with large synthesis simplicity.

At the final *System-Global Routing* level, the circuit is organized with a macro-regularity. Logic blocks defined at the previous level are regularly arranged and interconnected. The organization is generally an island-style scheme and the interconnect resources handle the resource limited interconnection pattern. This means that each logic blocks could be interconnected to others, but it is not possible to realize all the connections at the same time. In the FPGA scheme, we can also find the CLBs surrounded by the programmable interconnect. Several other interconnect patterns can be found with for example a local connection set between the logic blocks, such as those defined in [38].

This generic template represents a real opportunity for design exploration. Indeed, the denomination of a very generic template helps in enlarging each level. In fact, it allows extension of both technologies and architectures of each level, while keeping the others layer unchanged. This allows each layer to be studied independently, while the assumptions on all the other parts remain unchanged and greatly stimulate the development of original contributions.

2.4 Conclusion and Work Position

In this chapter, we surveyed the state-of-the-art regarding reconfigurable architectures. The FPGA structure is the traditional reconfigurable architecture. The basement of the original FPGA is the hierarchic and homogeneous arrangement of logic blocks. The architecture suffers from the programming circuit in area. Indeed, the computation part of the FPGA structure occupies only a small amount of area, conversely to routing structures. Today, the structure evolutes to heterogeneity. Thus, it will reach new application classes and improve the computation/routing ratio. The heterogeneity is found at design level. Adjunction of several specific logic blocks such as memory blocks and DSPs allows increasing the versatility and the computation performance. Also found at the technology level, heterogeneity leads to the use of flash/MOS co-integration to non-volatile store the configuration.

The standard architecture is optimized for CMOS. Emerging technologies could leads to new architectural paradigms for reconfigurable computation. Emerging reconfigurable architectures have been previously envisaged through two approaches: the use of density-increased devices (i.e. ultra-scale devices) and the use of functionality-enhanced devices (i.e. devices with new highly functionality within the same area). Globally, the use of emerging technology leads to the improvement of performance compared to CMOS structures. Nevertheless, their immaturity and the technological hurdles make these solutions prospective.

In this book, we will use the emerging technologies to improve the standard FPGA approach. Two approaches will be used. We will first focus on the improvement of peripheral circuitries, i.e. memories and routing resources (Chap. 3). These peripheral circuits are expected to improve the standard FPGA scheme by their direct use within the routing resources (Chap. 4). Then, we will break the logic block paradigms (Chap. 5) and propose a new seed for architectural organization (Chap. 6). In addition, we will have in mind two requirements: the proposition of a credible fabrication process and the architectural compatibility with existing FPGAs.

References

1. J. Birkner, Reduce random-logic complexity. Electron. Des. **26**(17), 98–105 (1978)
2. W. Carter, K. Duong, R.H. Freeman, H. Hsieh, J.Y. Ja, J.E. Mahoney, L.T. Ngo, S.L. Sze, A user programmable reconfigurable gate array, in *Procedings of the Custom Integrated Circuits Conference*, May 1986, pp. 233–235

3. H.S. Stone, Parallel processing with the perfect shuffle. IEEE Trans. Comp. C **20**(2), 153–161 (1971)
4. Altera Quartus-II: Altera FPGAs Design IDE, http://www.altera.com/products/software/sfw-index.jsp
5. Xilinx ISE: Xilinx FPGAs Design IDE, http://www.xilinx.com/products/design-tools/ise-design-suite/index.htm
6. Verilog-To-Routing (VTR) Project, http://www.eecg.utoronto.ca/vtr/
7. D. Marple, L. Cooke, An MPGA compatible FPGA architecture. *Proceedings of the IEEE Custom Integrated Circuits Conference*, 3–6 May 1992, pp. 4.2.1–4.2.4
8. P. Dillien, Adaptive hardware becomes a reality using electrically reconfigurable arrays (ERAs), in *IEE Colloquium on User-Configurable Logic—Technology and Applications*, 1 Mar 1991, pp. 2/1–2/10
9. A. El Gamal, J. Greene, J. Reyneri, E. Rogoyski, K.A. El-Ayat, A. Mohsen, An architecture for electrically configurable gate arrays. IEEE J. Solid-State Circuits **24**(2), 394–398 (1989)
10. B. Small, The flexibility of the quicklogic FPGA architecture. WESCON, 27–29, pp. 688–691 (1994)
11. V. Betz, J. Rose, A. Marquart, *Architecture and CAD for deep-submicron FPGAs* (Kluwer Academic Publishers, New York, 1999)
12. J. Rose, S. Brown, Flexibility of interconnection structures for field-programmable gate arrays. IEEE J. Solid State Circuits **26**(3), 277–282 (1991)
13. E. Ahmed, The effect of logic block granularity on deep-submicron FPGA performance and density. Master thesis, University of Toronto (2001)
14. E. Ahmed, J. Rose, The effect of LUT and cluster size on deep-submicron FPGA performance and density. IEEE Trans. Very Large Scale Integr. (VLSI) Syst. **12**(3), 288–298, (2004)
15. Xilinx Virtex-6 FPGA Family Overview (2011), http://www.xilinx.com/support/documentation/data_sheets/ds150.pdf.Accessed 24 March 2011
16. Xilinx Spartan-3A FPGA Product Overview (2011), http://www.xilinx.com/support/documentation/data_sheets/ds557.pdf.Accessed 4 April 2011
17. J. McCollum, H.-S. Chen, F. Hawley, Non-volatile programmable memory cell for programmable logic array, U.S. Patent No. 0064484, 2007
18. J. McCollum, G. Bakker, J. Greene, Non-volatile look-up table for an FPGA, U.S. Patent No. 0007293, 2008
19. J. Lipp, D. Freeman, U. Broze, M. Caywood, G. Nolan, A general purpose, non-volatile reprogrammable switch, WO Patent No. 01499, 1996
20. K.J. Han, N. Chan, S. Kim, B. Leung, V. Hecht, B. Cronquist, A novel flash-based FPGA technology with deep trench isolation. *Proceedings of the IEEE Non-Volatile Semiconductor Memory Workshop*, 26–30 Aug. 2007, pp. 32–33
21. N. Bruchon, L. Torres, G. Sassatelli, G. Cambon, New nonvolatile FPGA concept using magnetic tunneling junction, in *Proceedings of the IEEE Computer Society Annual Symposium on Emerging VLSI Technologies and Architectures*, 2–3 March 2006, p. 6
22. M. Lin, A. El Gamal, Y.-C. Lu, S. Wong, Performance benefits of monolithically stacked 3-D FPGA. IEEE Trans. Comput. Aided Des. Integr. Circuits Syst. **26**(2), 216–229 (2007)
23. M. Hutton, V. Chan, P. Kazarian, V. Maruri, T. Ngai, J. Park, R. Patel, B. Pedersen, J. Schleicher, S. Shumarayev, Interconnect enhancements for a high-speed PLD architecture, in *Proceedings of the ACM/SIGDA 10th International Symposium on FPGA* (2002), p. 3
24. D. Lewis, E. Ahmed, G. Baeckler, V. Betz, M. Bourgeault, D. Cashman, D. Galloway, M. Hutton, C. Lane, A. Lee, P. Leventis, S. Marquardt, C. McClintock, K. Padalia, B. Pedersen, G. Powell, B. Ratchev, S. Reddy, J. Schleicher, K. Stevens, R. Yuan, R. Cliff, J. Rose, The Stratix II logic and routing architecture, in *Abstracts of the ACM/SIGDA 13th International Symposium on. FPGA* (2005), p. 14
25. E. Kusse, J. Rabaey, Low-energy embedded FPGA structures Paper presented at the International Symposium on Low Power Electronics and Design (1998), p. 155

26. L. Shang, A.S. Kaviani, K. Bathala, Dynamic power consumption in Virtex-II FPGA family, in *Proceedings of the ACM/SIGDA tenth International Symposium on. FPGA* (2002), p. 157
27. V. Degalahal, T. Tuan, Methodology for high level estimation of FPGA power consumption, in *Proceedings of the Design Automation Conference*, 2005, p. 657
28. I. Kuon, J. Rose, Measuring the gap between FPGAs and ASICs, in *Proceedings of the ACM/SIGDA 14th International Symposium on FPGA*, 2006, p. 21
29. Emerging Research Devices and Materials Chapters, Updated Editions, International Technology Roadmap for Semiconductors (2010), http://www.itrs.net/Links/2010ITRS/Home2010.htm
30. R.C. Eberhardt, R.W. Dobbins, *Neural Networks PC Tools—A practical guide*. (Academic Press Inc., San Diego, 1990)
31. J.J. Hopfield, Neural networks and physical systems with emergent collective computational abilities. Proc. Natl. Acad. Sci. U.S.A. **79**, 2254–2258 (1982)
32. J. Hoekstra, E. Rouw, in *Modeling of Dendritic Computation: The Single Dendrite*, vol. 517 (The American Institute of Physics, Melville, 2000), pp. 308–322
33. C. Mead, *Analog VLSI and Neural Systems* (Addison Wesley, Menlo Park, 1989)
34. D.B. Strukov, K.K. Likharev, Reconfigurable hybrid CMOS/nanodevice circuits for image processing. IEEE Trans. Nanotechnol. **6**(6), 696–710 (2007)
35. W. Wu, G.-Y. Jung, D.L. Olynick, J. Straznicky, Z. Li, X. Li, D.A.A. Ohlberg, Y. Chen, S.-Y. Wang, J.A. Liddle, W.M. Tong, R.S. Williams, One-kilobit cross-bar molecular memory circuits at 30-nm half-pitch fabricated by nanoimprint lithography. Appl. Phys. A Mater. Sci. Process. **80**(6), 1173–1178 (2005)
36. Y. Luo, C.P. Collier, J.O. Jeppesen, K.A. Nielsen, E. Delonno, G. Ho, J. Perkins, H.-R. Tseng, T. Yamamoto, J.F. Stoddart, J.R. Heath, Two-dimensional molecular electronics circuits. J. Chem. Phys. Phys. Chem. **3**, 519–525 (2002)
37. J.E. Green, J. Wook Choi, A. Boukai, Y. Bunimovich, E. Johnston-Halperin, E. Deionno, Y. Luo, B.A. Sheriff, K. Xu, Y. Shik Shin, H.-R. Tseng, J.F. Stoddart, J.R. Heath, A 160-kilobit molecular electronic memory patterned at 1011 bits per square centimetre. Nature **445**, 414–417 (2007)
38. S.C. Goldstein, M. Budiu, NanoFabrics: spatial computing using molecular electronics, in *Proceedings of the 28th Annual International Symposium on Computer Architecture* (2001), pp. 178–189
39. S. Goldstein, D. Rosewater, Digital logic using molecular electronics.*Proceedings of the IEEE International Solid-State Circuits Conference*. vol. 1, 2002, pp. 204–459
40. A. DeHon, Array-based architecture for FET-based, nanoscale electronics. IEEE Trans. Nanotechnol. **2**(1), 23–32 (2003)
41. A. DeHon, M.J. Wilson, Nanowire-based sublithographic programmable logic arrays, in *Proceedings of the 2004 ACM/SIGDA 12th International Symposium on Field Programmable Gate Arrays* (2004)
42. A. DeHon, P. Lincoln, J. Savage, Stochastic assembly of sublithographic nanoscale interfaces. IEEE Trans. Nanotechnol. **2**(3), 165–174 (2003)
43. C.A. Moritz, T. Wang, Latching on the wire and pipelining in nanoscale designs, in *Proceedings of the 3rd Workshop on Non-Silicon Computation (NSC-3)*, June 2004
44. T. Wang, P. Narayanan, C. A. Moritz, Combining 2-level logic families in grid-based nanoscale fabrics, in *Proceedings of the IEEE/ACM International Symposium on Nanoscale Architectures*(NANOARCH), Oct 2007
45. P. Vijayakumar, P. Narayanan, I. Koren, C.M. Krishna, C.A. Moritz, Impact of nanomanufacturing flow on systematic yield losses in nanoscale fabrics, in *Proceedings of the IEEE/ACM International Symposium on Nanoscale Architectures* (NANOARCH), June 2011
46. H. Yan, H.S. Choe, S.W. Nam, Y. Hu, S. Das, J.F. Klemic, J.C. Ellenbogen, C.M. Lieber, Programmable nanowire circuits for nanoprocessors. Nature Lett. **470**, 240–244 (2011)
47. Y.-M. Lin, J. Appenzeller, J. Knoch, P. Avouris, High-performance carbon nanotube field-effect transistor with tunable polarities. IEEE Trans. Nanotechnol. **4**(5), 481–489 (2005)

48. M.H. Ben Jamaa, D. Atienza, Y. Leblebici, G. De Micheli, Programmable logic circuits based on ambipolar CNFET, in *Proceedings of the 45th ACM/IEEE Design Automation Conference*, 8–13 June 2008, pp. 339–340
49. M.H. Ben Jamaa, K. Mohanram, G. De Micheli, Novel library of logic gates with ambipolar CNTFETs: opportunities for multi-level logic synthesis. Paper presented at the Design, Automation & Test in Europe Conference & Exhibition, 20–24 April 2009, pp. 622–627
50. I. O'Connor, J. Liu, F. Gaffiot, F. Pregaldiny, C. Lallement, C. Maneux, J. Goguet, S. Fregonese, T. Zimmer, L. Anghel, T.-T. Dang, R. Leveugle, CNTFET modeling and reconfigurable logic-circuit design. IEEE Trans. Circuits Syst. I Regul. Pap. **54**(11), 2365–2379 (2007)
51. N. Patil, J. Deng, A. Lin, H.S.-P. Wong, S. Mitra, Design methods for misaligned and mis-positioned carbon-nanotube-immune circuits. IEEE Trans. Comp. Aided Des. (2008)
52. N. Patil, A. Lin, J. Zhang, H. Wei, K. Anderson, H.-S.P. Wong, S. Mitra, Scalable carbon nanotube computational and storage circuits immune to metallic and mis-positioned carbon nanotubes. IEEE Trans. Nanotechnol. (2010)
53. J.R. Heath, P.J. Kuekes, G.S. Snider R.S. Williams, A defect-tolerant computer architecture: opportunities for nanotechnology. Science **280**(5370), 1716–1721 (1998)
54. W. Huffman, V. Pless, *Fundamentals of Error-Correcting Codes* (Cambridge University Press, Cambridge, 2003)

Part II
Incremental Logic Design

Chapter 3
Innovative Structures for Routing and Configuration

Abstract The goal of this chapter is to illustrate how emerging technologies can help to improve performance metrics of conventional Field-Programmable Gate Arrays structures. It is widely recognized that in traditional FPGAs, both the memory and the routing circuitry (with 43% of area for each contribution) represent the principal bottleneck to scaling and performance increase. In this context, we investigated 3D integration techniques for passive and active devices. The technologies surveyed will be a resistive memory technology, monolithic 3D integration and a vertical 1D transistor technology.

In the previous chapter, we have seen that most area and performance metrics of modern Field Programmable Gate Array are limited by configurable interconnect and configuration memories. In this chapter, we will see how emerging technologies might be used to improve these two aspects. 3D integration techniques will be surveyed. We will first discuss the exact context, in order to define the objectives of the designed blocks in a clear way. The various technologies surveyed are a resistive memory technology, monolithic 3D integration and a vertical 1D transistor technology. Each technology will be addressed with the following organization. After a brief literature review, we will describe the technology assumptions. Then, novel logic blocks will be presented and characterized.

3.1 Context and Objectives

3.1.1 Context Position

In Sect. 2.1.3, the distribution of the area occupation between logic, memory and routing resources within an FPGA has been described. It shows that almost 45% of the silicon area is used only for the configuration memories, while the routing

P.-E. Gaillardon et al., *Disruptive Logic Architectures and Technologies*,
DOI: 10.1007/978-1-4614-3058-2_3,

Table 3.1 Specification estimation wrt. memory distribution through an FPGA (extracted from Xilinx Virtex6 architecture [1])

	Registers	LUTs	Connection boxes	Switchboxes	Memory blocks
Read operations	10^8 s^{-1}	10^8 s^{-1}	10	10	10^7 s^{-1}
Write operations	10^8 s^{-1}	10	10	10	10^7 s^{-1}
Number	1×10^6	30.3×10^6	20×10^6	20×10^6	38×10^6
Distribution	Even distribution throughout the architecture (fine-grain blocks)				Standalone blocks

resources occupy 78% of the total area. In this chapter, we will address these two specific parts of the FPGA: memory and the routing resources.

The requirement for storage in an FPGA could be considered at several levels. This distribution means that designers face various constraints. Table 3.1 shows a study on memory requirements in an FPGA and gives an overview of associated constraints. We consider three different types of storage: high speed-data storage (fast flip-flops in the BLEs); configuration memories for logic (LUTs) and routing (Connection boxes and Switchboxes); and standalone Random Access Memories (Dedicated RAM Blocks). We focus on configuration memories (LUTs, CBs and SBs). These memories represent the biggest part of the FPGA area and share the same properties. They are distributed throughout the logic circuit and are programmed only a small amount of time. Traditionally, they are realized by SRAM circuits. SRAMs ensure a technical homogeneity between the logic and the configuration part.

Nevertheless, SRAM memories are power- and area-consuming, as well as volatile circuits. This means that the configuration must be loaded at each power-up, wasting time and power. Several circuits proposed the use of flash memories to create a non-volatile configuration [2]. However, the flash technology has a long programming time, and requires process co-integration. Indeed, floating-gate transistor processes require more steps than high-performance CMOS processes. This obviously incurs extra fabrication costs and technical difficulties. Such a non-volatile technology is thus adapted only to niche applications. Hence, the technology of configuration memories should be improved by using low-cost and non-volatile technologies.

Furthermore, the largest part of the configuration memories are used to configure the routing circuits (82% of the memory area). The reconfigurable interconnect alone occupies 45% of the area and introduces many active devices within the data paths. These devices impact directly the performance metrics of the structure, by increasing the critical path delay of the reconfigurable architecture. Thus, it is of great interest to consider the problem of "routing" in a global way by addressing the question of memory and the question of active configured devices at the same time. Hence, while memory and routing could be addressed separately, it is reasonable to work on both sides try to compact this entire periphery.

3.1.2 Objectives

In this chapter, we will propose the use of 3D techniques, to place devices in the back-end layers. Three different technologies will be surveyed. The difference resides in the type of device that is placed in the back-end.

Firstly, Phase-Change Memories (PCM) will be used to embed a passive resistive memory above the IC. Such a device is non-volatile and technologically compatible (i.e. indicating homogeneous integration) with the CMOS technological process. Hence, we will propose an elementary memory node, able to store a configuration in the resistive state of the memory and to provide it intrinsically to a logic gate. The storage of a configuration for a resistive memory is quite obvious. Nevertheless, such a technology is of high interest due to the low on-resistance characteristic of the memory. This makes it possible to directly use a memory as a high-performance switch and to embed it directly within the logic data paths. We will thus propose a switchbox circuit that uses simple resistive elements to replace SRAMs and routing pass gates. All these circuits will be compared to their elementary CMOS FPGA counterparts.

Secondly, we will use monolithic 3D integration technology to stack active devices with a high via density. Such a process allows the stacking of 2D active devices. We propose to split the configuration memories and the data path transistors. This allows technological improvement of both classes of circuit. We suggest a complete integration of simple FPGA blocks, such as configuration memory, LUT and pass-gates, down to the layout level. In this way, we can provide a performance evaluation of elementary nodes with regards to standard CMOS FPGAs.

The previous technology proposes an integration of devices in a 3D manner. Nevertheless, only 2D devices are stacked. We therefore propose an integration process, which aims to realize a transistor (channel) in the vertical direction. A vertical NanoWire Field Effect Transistor process allows a vertical orientation of the active part of the transistor. It is then envisaged that several routing circuitries (such as programmable vias and signal buffers) can be embedded in the back-end layer. In order to evaluate the performance metrics of the technology and compare it to that of CMOS, we present a methodology based on TCAD simulations. TCAD will be used to model the elementary device, and electrical simulations of simple circuits will be performed.

Finally, we will draw a global comparison between the technologies and extract some overall conclusions.

3.2 On the Use of Resistive Memory Technologies

While Static Random Access Memories, Dynamic Random Access Memories (DRAMs) and Flash memories are predominant in microelectronics systems, thanks to their CMOS process compatibility, a large number of new memory devices have been highlighted by the International Technology Roadmap for Semiconductors [3]. These memories are generally based on new physical phenomena to retain the information and lift roadblocks to high density integration. In this sub-chapter, we will focus on non-volatile resistive memories.

3.2.1 Introduction

Next-generation *Non-Volatile Memory* (NVM) has attracted extensive attention due to conventional memories approaching their scaling limits. Several types of NVMs, such as ferroelectric random access memory, magnetic random access memory, and Resistive Random Access Memory (ReRAM), are being investigated. Among various NVMs, ReRAMs are typically composed of a simple metal-switching element-metal structure, which has the merits of low power consumption, high-speed operation, high-density integration and CMOS process compatibility.

Resistive memories, which can see their resistance vary depending on the applied voltage, were intensively studied from the 1960s to early 1980s for device applications [4]. Several materials can be envisaged to execute this desired functionality. The type of material determines the physical phenomena that are used in the resistive change. For example, chalcogenide materials, semiconductors, various kinds of oxides and nitrides, and even organic materials were found to have resistive memory properties. Hence, the architecture will depend to a large extent on the technology. We can classify the technologies into two main families: Oxide Memories (OxM) and Phase-Change Memories.

Oxide memory state change is accomplished by the creation or the destruction of a conductive bridge through an oxide layer. This property is due to different physical phenomena, which depend on the material. The conduction forming mechanism is still not fully understood, and is currently under investigation. The structure is composed of a changeable resistance material sandwiched between two terminal electrodes. Resistance change can be achieved by controlling the current or voltage pulse applied to the electrodes, and the resistance state remains stable without being refreshed. To date, a number of different switching characteristics have been observed in a variety of material systems; including NiO_2 [5, 6], TiO_2 [7], HfO_2 [8], WO_x [9], CuO_x [10], TaO_x [11]. In fact, it has become well understood that a number of combinations of an oxide with metal electrodes can exhibit some kind of resistance switching behavior.

Phase-Change Memories are, as the name indicates, based on a material having two different stable physical phases leading to two different resistances. As with O_xM, several materials might be used such as GeSbTe [12], GeTe [13], GeTeC [14]. The PCMs are considered today to be one of the most promising candidates for the next generation of non-volatile memory applications [15]. The interest in PCMs is due to various advantages, including: better scalability (down to a few nanometers) [16], faster programming time (of the order of few nanoseconds) [17] and improved endurance (up to 10^9 programming cycles) [18]. Some prototypes (such as a 60-nm 512-Mb [19] and a 45-nm 1-Gb [18] PCM technology) have been presented recently to showcase the viability of high-density standalone memories based on PCM technology from an industrial point of view. The PCM technology achieves the maturity required for large applications. We will focus on this technology. Nevertheless, it is worth pointing out that this work can be generalized to any other resistive memory technology.

3.2.2 Phase Change Memory Properties and Technological Assumptions

3.2.2.1 Physical Phenomena

A PCM device is based on the electrothermal-induced reversible phase transition of a chalcogenide alloy between an amorphous insulating state (RESET) and a polycrystalline conductive state (SET). The polycrystalline phase is inherently stable, as it is the lowest possible energy state of the system. On the other hand, retention instability affects the amorphous phase through two physical phenomena: spontaneous crystallization and low-field conductivity drift [20]. Recently, many efforts have been made in order to achieve a better understanding of the physical mechanisms which govern the behaviour of amorphous chalcogenides integrated in PCM cells [21–24]. Chalcogenide alloys are semiconducting glasses made by elements of the VI group of the periodic table, such as sulphur, selenium and tellurium. The best-known and most widely used chalcogenide alloy is $Ge_2Sb_2Te_5$ (GST). GST can guarantee stability of programmed amorphous bits for more than 10 years at 85°C [25]. While this can be considered to be sufficient for consumer applications, many efforts are today devoted to the development of new chalcogenide materials to improve the high-temperature reliability of PCM technologies in order to address the embedded memory market as well. Recent findings show that GeTe thin films demonstrate a higher crystallization temperature than GST [12], as well as superior data retention performances when integrated in memory cells [13]. Furthermore, it is known that the crystallization process is affected by the presence of foreign atoms in the material. For example it has been demonstrated that the Nitrogen doping of GST dramatically increases the stability of amorphous phase [26, 27].

Indeed, by means of a careful control of Joule heating through the cell, it is possible to electrically switch the chalcogenide layer between its two stable configurations, i.e. the high-conductive polycrystalline state and a low-conductive amorphous one, as shown in Fig. 3.1. A sufficiently high voltage pulses heat into the Phase-Change (PC) layer above the melting temperature of the material (T_m). A rapid quench follows and part of the chalcogenide alloy (depicted as an oval in the PC layer) is stuck in the amorphous phase. The resulting memory cell is in a

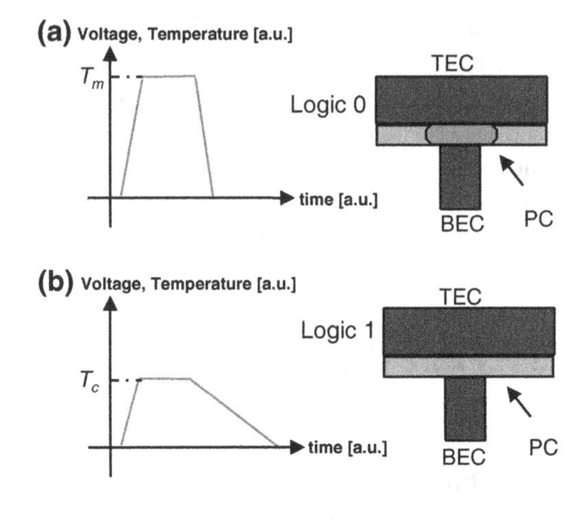

Fig. 3.1 Schematic of PCM device in Logic 0 (named RESET) and Logic 1 (named SET) configurations and of the programming pulses suitable to obtain the states

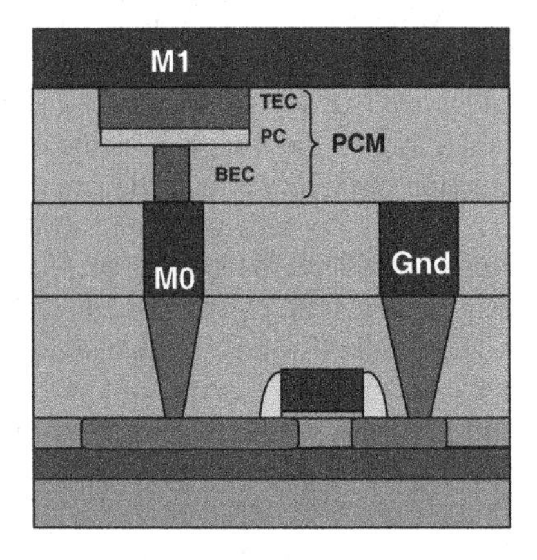

Fig. 3.2 Cross sectional schematic showing a PCM device integration

high resistance state (Fig. 3.1a). A lower but longer pulse is used to crystallize the amorphous region of the PC layer in order to achieve a low resistance memory cell (Fig. 3.1b).

3.2.2.2 Technological Assumptions

PCM technology is CMOS-compatible. As in Flash-NOR arrays, each memory cell includes a storage phase-change node and a selector transistor in series (i.e. *1-resistor-1-transistor* configuration). The memory element may be fabricated either just after the Si contact forming step at the Front-End-Of-Line (FEOL) level or after the first steps of interconnections at the Back-End-Of-Line (BEOL) level, (e.g. on top of the Metal 0 or Metal 1 interconnect level) [28]. A schematic cross-section of the storage element architecture is shown in Fig. 3.2. The PCM device, formed of a PC layer with Bottom (BEC) and Top (TEC) Electrode Contacts, is integrated between M0 and M1 interconnection level in the back-end-of-line. The MOSFET selector (bottom) is fabricated in the front-end-of-line. This figure depicts a pillar structure. The pillar approach is the simplest way to create a PCM device. First, a metallic heater is built. The heater is made by etching a via into the inter-layer dielectric and by filling it with a metal. The role of the heater is to help to channel the current in order to increase its density and thus maximize the heat control in the memory node. To improve the heater fabrication, several sublithographic techniques have been proposed [29, 30]. After the heater metal deposition, the via is filled by chalcogenide alloys with a room temperature deposition. The top electrode is obtained by a final metal deposition.

3.2.2.3 Opportunities

Resistive memories, and especially the envisaged Phase-Change technology, represent truly promising opportunities for several aspects of design. Indeed, PCMs demonstrates non-volatile behavior at low cost. Such a property is obviously of high interest for all types of reconfigurable circuits, where a permanent configuration circuit is strongly desirable. Furthermore, the technology is fully compatible with Back-End of Line and able to integrate the memories into the 3rd dimension. This makes the resistive memories highly relevant as configuration points, since we could expect promising size reduction through the integration, above active silicon, of all area-hungry memories. Furthermore, the resistance of the on-state is typically below 1 kΩ (for example that of GeTe is around 50 Ω). This is far lower than any MOS switch. Thus, it makes sense to use them as a high performance switch replacement element for FPGAs, by directly introducing them into the logic data path. In such a way, we expect not only to improve the size of the routing elements, but also to drastically reduce the delay of implemented circuits.

3.2.3 Elementary Memory Node

In this section, we present an elementary circuit, based on a PCM non-volatile resistive memory, used to move most of the configuration part of reprogrammable circuits to the back-end, reducing their impact on front-end occupancy. Such a memory node is dedicated to drive multiplexer (MUX) inputs or pass-gates. The memory node is programmed by injecting a certain current through it; while the information has to be read as a voltage level. Furthermore, it shall allow a layout-efficient line sharing.

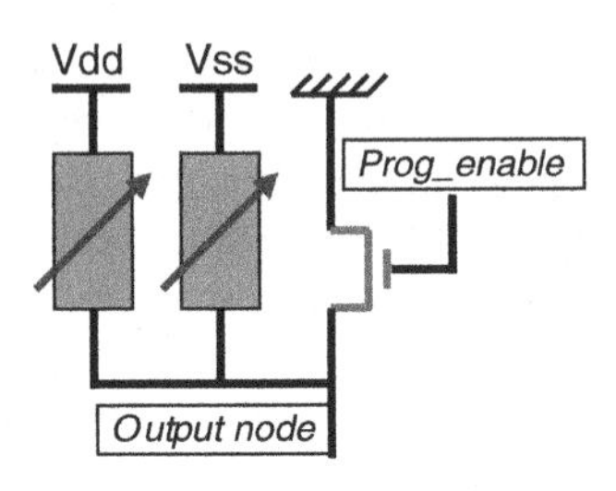

Fig. 3.3 Logic-in-PCM elementary memory node

3.2.3.1 Concept

The elementary memory node is presented in Fig. 3.3. The circuit consists of two resistive memory nodes connected in a voltage divider configuration between two fixed voltage lines. A transistor is also connected between the ground and the output node of the cell. It is used to select the node during the programming phase. The output is designed to place a fixed voltage on a classical standard cell input. Read operations are intrinsic to the structure, while programming is an external operation to be performed on the cell.

The voltage divider is used in this topology to execute intrinsically the conversion from a data stored in a variable resistance to a voltage signal. Figure 3.4 shows a configuration example where the node stores a '1'. The programming transistor is placed in the off-state by the non-active *Prog_Enable* signal, so that the ground is disconnected from the output. The resistive memory (1) that is connected to the V_{dd} line, is configured into the crystalline state, so its associated resistivity is low (a few kΩ). The other memory (2), connected to V_{ss}, is in the amorphous state with high resistivity (close to 1 MΩ). As a consequence, a voltage divider is configured and the output node is charged close to the voltage of the branch with a high conductivity. The logic levels depend on R_{ON} and R_{OFF} as in the following relations:

$$'1' = Vdd - \frac{R_{ON}}{R_{ON} + R_{OFF}}(Vdd - Vss) \quad '0' = \frac{R_{ON}}{R_{ON} + R_{OFF}}(Vdd - Vss)$$

It is also worth noticing that in continuous read operation, a current will be established through the resistors. This leads to a passive current consumption through the structure depending to the following relation:

$$I = \frac{Vdd - Vss}{R_{ON} + R_{OFF}} \approx \frac{Vdd - Vss}{R_{OFF}}$$

As an illustration, the PCM technology off-resistance is around the MΩ value. At $V_{dd} = 1$ V, the technology yields to a leakage current of 1 μA. This is obviously too high for a viable industrial solution. Nevertheless, this static current could be reduced by the choice of a memory technology maximizing the R_{OFF} value (e.g. OxRAM technology exhibits off-resistance bigger than the GΩ).

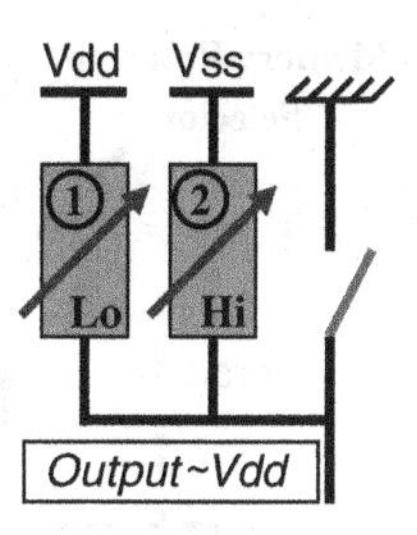

Fig. 3.4 Node in read configuration

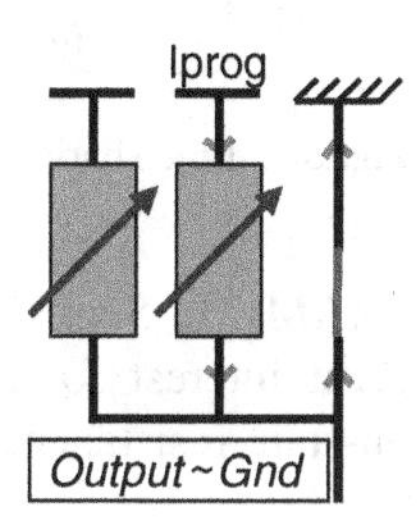

Fig. 3.5 Node in write configuration

3.2.3.2 Programming Circuitry

Figure 3.5 presents the programming phase of the node. While the programming transistor is placed in the on-state by setting the *Prog_enable* signal, the fixed read voltage sources are disconnected from the top lines and replaced by the programming unit. Then, a programming current is applied sequentially into the resistive memories to change their states. Programming currents are drained to the ground.

As each cell has its own selection transistor, the programming lines can be shared in a standalone-memory-like architecture, as shown in Fig. 3.6. The programming unit is composed of three different elements. A programming pulse generator handles the creation of the programming pulse with the correct waveforms. The programming signals are then routed by two stages of multiplexers. They are routed through BL_{XA} or BL_{XB} lines in order to program memory A or B respectively. We also note the Program/Operation selectors. Their aim is to route static voltages when the nodes are not under programming. During the programming, the selection of a node is ensured by the WL_X signals. Thus, the choice of the memory node to program is made at the programming unit level, through the selection realized by the Program/ Operation selectors and the Memory Program selector.

3.2.4 Routing Elements

In the previous section, we presented a memory node, which aims to replace the configuration memories in configuration memory-intensive circuits. The node is intended to drive any logic gate and is thus a straightforward replacement part for

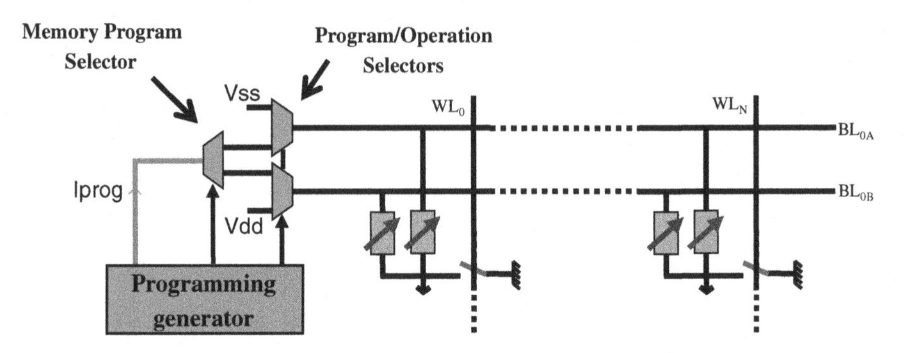

Fig. 3.6 Line sharing illustration in standalone-memory-like architecture

SRAMs. In Sect. 3.2.2.3, we observed that the use of Resistive RAMs is also of great interest for performance improvement of data paths, thanks to their low on-state resistance.

3.2.4.1 Concept

FPGA switchboxes represent the most demanding circuits in terms of routing performance. We thus propose to replace the traditional pass gates with the structure shown in Fig. 3.7. Figure 3.7a shows a schematic diagram of a 2×2 crossbar structure using PCM resistances. At each cross point in the crossbar, routing elements are placed. They are able to create any combination of connections between the North/South/East/West terminals of the cross point. These routing elements are built with a similar structure as those of CMOS, whereby two-terminal PCMs replace pass-transistors (Fig. 3.7b). An on-connection is realized by programming the PCM, situated between the two wires that it should connect, to a low resistance state. With this structure, we build an "intelligent cross point", which is able to merge the programming node and the pass switch in a single device. The device is embedded in the Back-End levels as a via, and replaces a five-transistor SRAM and a four-transistor pass gate. Furthermore, since the PCM itself performs the switching in the data path, its own on- and off-resistance values also have a direct impact on the circuit performance.

Once programmed, the cell behavior is purely static. Conductive paths are created by the resistive networks as illustrated in Fig. 3.7c. In this figure, we observe three conductive paths through the box: InW →OutN, InN→ OutE, and InE →OutW. In this structure, the number of reachable inputs/outputs depends on the R_{on}/R_{off} ratio. In such crossbar architecture, the lowest resistive memory paths define the connections. Nevertheless, currents will also be established in the high conductive bridges. Thus, the discrimination between a conductive and a non-conductive path rely only on the path resistance difference. We consider the situation of Fig. 3.7c. The path InE → OutW goes through two on-resistances. This

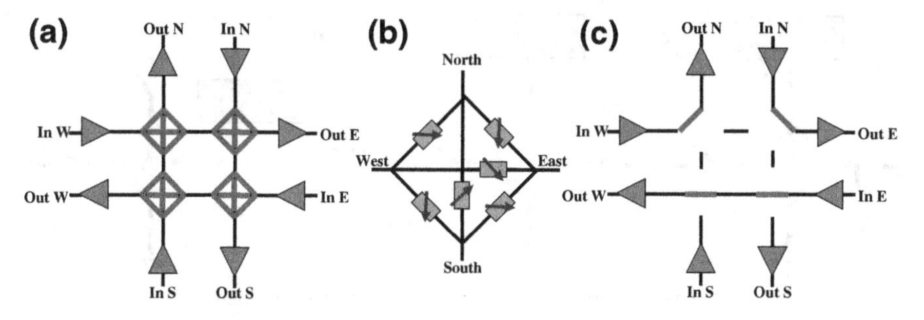

Fig. 3.7 **a** PCM-based 2 × 2 switchbox architecture **b** zoom on the cross point structure **c** example of programmed switchbox (memories in off-state are not shown for clarity)

resistance is thus 2 R_{on}. The path InE → OutS goes through a unique off-resistance. The resistance is thus R_{off}. In this example, it is not possible to discriminate the path if R_{on}/R_{off} is too low (i.e. two in the extreme bound). More precisely, the longest on-resistance path of the structure should be compared to the shortest off-resistive path. The tolerable R_{on}/R_{off} ratio is thus given by:

$$\frac{Longest\ Path\ R_{on}}{Shortest\ R_{off}} \geq 1000$$

(with 1000 an arbitrary choice to ensure a significant discrimination). More considerations on the structure can be found in [31].

3.2.4.2 Programming Circuitry

A PCM is programmed by applying a pulsed signal between its two terminals. This conducts the PCMs of the switchbox to be addressed sequentially. In the structure, the PCM is selected by connecting one terminal to the programming unit, while the other is grounded. When a unique memory is select, the other PCM terminals are left floating to avoid parasitic programming. After the selection, the programming unit drives the desired set or reset pulse to program the resistivity state. An example of sequential programming is shown in Fig. 3.8.

The programming voltages and timing pulses required to program the PCMs may be applied through the drivers at the inputs and the outputs of the switchboxes (Fig. 3.7a). Figure 3.9 shows a possible structure for these drivers. They are used for the electrical interface between signal channels and the programming unit, which generates the configuration waveforms. Figure 3.9a shows a possible input driver, while Fig. 3.9b shows the implementation of the output driver. As explained, the buffers must allow the connection of the nodes to the programming unit, to the circuit or to a high-impedance node. This last possibility is handled by the three-state multiplexers and buffers. The programming unit is routed by the multiplexer for the input, and through a single pass transistor for the output.

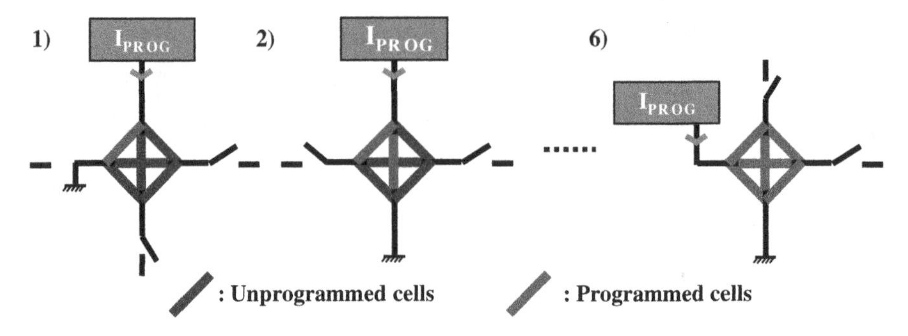

Fig. 3.8 Example of switchbox programming sequence

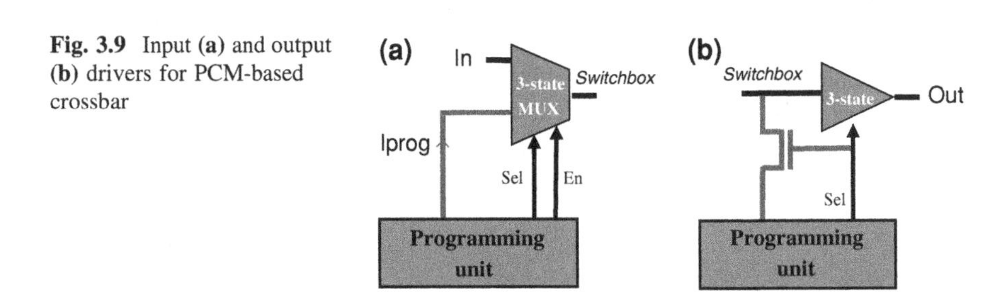

Fig. 3.9 Input (**a**) and output (**b**) drivers for PCM-based crossbar

3.2.5 Performance Characterization

3.2.5.1 Methodology

In order to characterize the performance of the node, we extracted the area, the write time and the programming energy as metrics. The extraction is based on node complexity and interpolation from ITRS figures [32]. Then, comparison with memory elements traditionally used in FPGA, such as MOS SRAM 5T [33] and flash memories storage LUT elements [34], are used to benchmark the structure.

Since the technology enables most of the configuration circuit to be placed in the back-end levels, the area metric corresponds to the front-end projection of the impact of the memory node. The write time and the programming energy are also considered. Even if the use of non-volatility makes these considerations far less critical, it is still of interest to improve them, in order to enable low power and fast reprogramming.

3.2.5.2 Performance Estimation

Table 3.2 shows some characterization results in terms of area and write time for the proposed solution compared to traditional FPGA memory nodes. Considering

Table 3.2 Detailed technology performance evaluation

	Cell elements	Area (F^2)	Write time (ns) [37]	Programming energy (pJ) [37]
SRAM	5T	115	0.3	7.10^{-4}
Flash cell	2T	46	1000	0.1
MRAM UFF	5T2R	115	40	300
PCM cell	1T2R	30	60	12
Flash versus PCM	–	× 1.5	× 16.6	× 0.08
MRAM versus PCM	–	× 3.8	× 0.6	× 25

that all these elements are driving an equal load, we omitted the pass transistor at the output node in our simulations. Thus, the SRAM cell is considered to be a 5-transistor (5T) structure. The Flash topology is implemented by 2 Flash transistors [35]. The Magnetic RAM implementation is realized by an unbalanced flip flop as proposed in [36]. In fact, this allows efficient separation of the programming path from the data path.

We see that the proposed PCM cell is the most compact solution, even with the impact of the programming current on the access transistor. This advantage is due to the reduction of the memory front-end footprint to only one transistor, compared to five for the SRAM cell, and compared to two for the Flash solution (one pull-up transistor coupled to a floating gate transistor). We should remark that PCMs offer a significant reduction in writing time for non-volatile memory technologies, as well as a consequent reduction in writing energy. In our context, it is possible to reduce the area by a factor of 1.5 and the writing time by a factor of 16.6 and the programming energy by a factor of 500 compared to an equivalent flash technology. However, the programming energy is about ten times larger than in the flash technology. Compared to another emerging equivalent non-volatile magnetic resistive technology, we can observe that the proposed structure improves the area by a factor of 3.8. This is due to the chosen structure for the magnetic resistive node, which is based on an area-hungry unbalanced flip-flop. Nevertheless, it is important to highlight that the PCM-based node is slightly slower for reconfiguration, with a 33% difference as compared to magnetic memory. Finally, while the writing time is slower, the required programming energy is reduced by a factor of 50 as compared to the magnetic technology, which requires energy to create the programming magnetic field. We should therefore consider that this result is strongly dependent on the technology used. In this work, we considered Thermally-Assisted-Switching (TAS) MRAM technology. While this technology has the same level of maturity as PCM, it also requires a large programming energy (between 100 and 150 pJ per cell). Most advanced writing schemes, such as Spin Transfer Torque [38, 39] allow further reduction of writing energy to the range of 3 pJ per cell.

Table 3.3 shows some characterization results in terms of area and writing time for the solution and conventional FPGA memory nodes. The SRAM-based node is

Table 3.3 Technology performance evaluation (2 × 2 switchbox)

	Cell elements	Area (F^2)	Write time (ns) [37]	On data path resistance (Ω)	Off data path resistance (MΩ)
SRAM	72T	2576	2.4	9100	144
Flash cell	48T	1104	24000	9100	144
PCM cell	8T24R	321	1440	50	1
MRAM UFF	168T48R	4448	960	9100	144
Flash versus PCM	–	× 3.4	× 16.6	× 182	× 0.007
MRAM versus PCM	–	× 14	× 0.6	× 182	× 0.007

formed by four 4-input multiplexers, programmed by 5T SRAMs. The Flash-based solution is taken from [35], while the MRAM-based solution is taken from [36]. In this work, two flash transistors are used to replace a combination of one SRAM and one pass gate. The structure of the switchbox is considered to be the envisaged PCM structure. We should note that a GeTe technology [14] with a very low on-resistance value is envisaged, as it corresponds to the requirements for a high performance switch.

We see that the proposed PCM switchbox is the most compact solution, even with the impact of the programming current on the access transistor. This advantage is due to the reduction of the memory front-end footprint to only metal lines, compared to five transistors for the SRAM cell and two transistors for the Flash solution (one pull-up transistor coupled to a floating gate transistor). It is also worth noticing that PCMs offer a significant reduction in write time versus flash technology. In our context, it is possible to reduce the area by 3.4 and the write time by 16.6 as compared to an equivalent Flash technology. Another important metric is the data path resistance. In the context of a logic signal propagating through the switchbox, we observe that the on-resistance of PCM is very low compared to the other solutions. In fact, all the other solutions still use a MOS transistor to realize the data path switch. Hence, since the on-resistance is decreased by a factor of 182, this means that a reduction of the propagation time is expected. From the point of view of the off-resistance however, we observe the opposite, since the off-resistance follows the same trend as the on-resistance. This means that the off-resistance decreases by a factor of 144, and since off-resistance impacts directly on switch leakage, this decrease is likely to worsen static power dissipation and logic levels. However, this could be mitigated by the fact that the R_{on}/R_{off} ratio has not decreased—on the contrary, PCM cells demonstrate an improved ratio with a value of around 1/20,000, with respect to the value of 1/15 824 obtained for the other solutions. The ideal solution will require a low R_{on} value, while the R_{on}/R_{off} ratio should be increased with respect to CMOS.

3.2.5.3 Discussion

The presented performance metrics make the solutions built around resistive memories of high interesting for the purposes of reconfigurable applications. Indeed, we showed that it is possible to create a compact configuration memory node that can store a logic level in two resistances. This logic node improves the size as compared to flash memories by 1.5×. While this 33% reduction in area reduction is significant, it is worth noticing that the limitation is due to the programming transistor. Indeed, the selection transistor size is dictated by the level of current to be driven to the cell being programmed. While the required programming current is large, the programming energy remains the lowest over the benched technologies. Other technologies might be envisaged in order to further reduce the programming current. First of all, other PCM technologies, such as GeTe or GeTeC [14], require smaller currents for programming compared to standard GST. It is then possible to migrate to other resistive memory technologies. For example, OxRAM requires less current overall for their programming, and will lead to more compact configuration nodes [40].

Concerning the routing part, we propose to introduce the PCM directly into the logic data path. This is of interest from two points of view. First, we obtain an area reduction of a factor of 3.4 as compared to flash. However, the most interesting advantage comes from the technology: the ReRAM technology is shown to have a much lower on-resistance than CMOS switches, which should lead to significant reductions in propagation delay in complex logic circuits. While the well-known GeSbTe-based PCM technology has a "quite high" R_{on} value, it is attractive to look at new PCM materials or new technologies. In particular, special alloys such as GeTeC and GeTe [14] have demonstrated a very small on-value, well-suited to the requirements of the routing structures. Nevertheless, it will be necessary also to study the reliability of the proposed switchboxes as well as the data retention time in the structure. Resistive memories are switched by a controlled current or voltage applied to or through them. Nevertheless, the switching behavior is generally more complex, as far as timing and environment (e.g. temperature) must also be considered. Unpredictable signals are flowing in the data path. Unpredictable means that the signals depend on the application. Thus, the memories introduced in the flow maybe switched to unwanted states leading to an unreliable structure.

3.3 On the Use of Monolithic 3D Integration Process

In the previous section, we looked at a technology, which allows a passive resistive memory device to be embedded into the back-end levels. While this approach is advantageous for memory or routing structures, the elements are only passive. It then appears interesting to assess the interest of 3D integration, in order to diversify the functionality of above IC devices. In this section, we propose the use

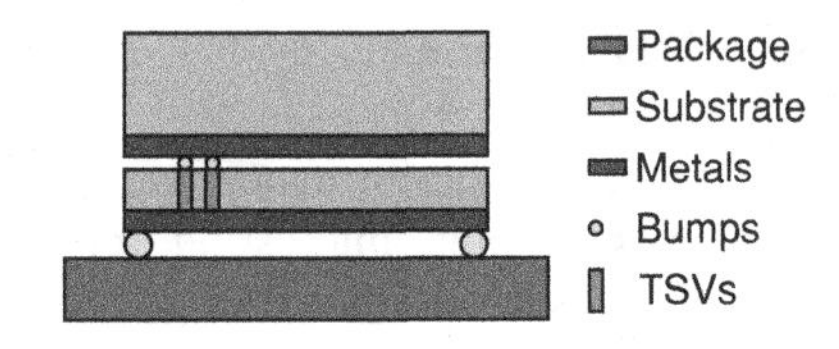

Fig. 3.10 TSVs-based die stacking cross-sectional view

of a monolithic 3D integration technology to stack two active layers. This represents a first step towards the use of active devices in a 3D scheme.

3.3.1 Introduction

For several decades, the semiconductor industry has invented new approaches to increase integration density and transistor performance, the main vector for this being MOSFET scaling.

In this context, the IC integration in three dimensions appears to be a promising alternative path to scaling, and to some extent would avoid the huge investments required by scaling. While the concept is not entirely new [41], the development of the technology has witnessed significant growth over the last decade. In particular, the technology of *Through-Silicon-Vias* (TSVs) is currently the reference for 3D technology processes. A TSV could be defined as a large via built across the substrate, in order to contact the active front-end to the reverse-side of the chip. Bumps are then used to contact the dies between them, as shown in Fig. 3.10.

While such integration is challenging due to TSV density requirements (leading to a double specification of TSV aspect ratio and wafer thickness), the process is mature enough for industrial applications [42]. However, there are still some hurdles to face. Table 3.4 depicts the principal characteristics of TSV processes [43]. It is worth noticing that TSVs are quite area-hungry with diameter up to 5 μm and pitch up to 10 μm. This means that the connection density is quite poor (from 400 to 10,000 TSVs/mm^2) and this results in a loss of active size. Thus, designers are limited to high level interconnect, such as memory/core communication. However, this segregation is of great interest for performance, since processes will be tuned to optimize each layer to a given class of application (low power, general purpose).

In a reconfigurable application, a large number of interconnections are required if separation between memory and logic is to be envisaged. This means that other integration processes should be used, to overcome the limitations of connection density. In an FPGA, connections between memory and logic are done at gate level. We estimate the required density at about 500,000 3D contacts/mm^2. Thus, instead of processing the layers separately and stacking them *a posteriori*, it is possible to use monolithic sequential integration. In a monolithic integration, the circuit is processed from the bottom to the top. This

Table 3.4 Density survey of TSV technologies [43]

	Diameter (μm)	Pitch (μm)	Density (TSVs/ mm^2)	CMOS 65-nm gate equivalent
Low density TSVs	20	50	400	1,250
High density TSVs	5	10	10,000	50

means that the stacked layers are build by a set of technological steps above the already processed stack.

Such an integration scheme is promising for several reasons. Firstly, it enables a much higher via integration density. In fact, the contacts use a planar scheme step, as opposed to TSVs. Thus, it is possible to obtain alignment accuracy in the range of lithographical precision (around 10-nm), whereas TSV alignment accuracy is in the range of 1 μm. However, monolithic integration gives rise to other issues. In particular, conversely to parallel integration, where the layers are processed separately, it is necessary for the fabrication of the top transistors not to degrade the performance of the bottom layer transistors. This means that the top layer thermal budget is limited and that processes have to be thought through in consequence.

3.3.2 Technological Assumptions

As briefly introduced above, the monolithic 3D integration process integrates at least two layers of active silicon. These layers are processed sequentially, as shown in Fig. 3.11. The principal process steps are (a) the realization of the bottom transistor (b) the deposit of the top film silicon and (c) the realization of the top transistor. Several constraints appear at each step.

The first transistor layer must be optimized in order to improve its robustness to temperature. It is worth nothing that the second active transistor layer will have an impact on the first layer and thus the thermal robustness of the bottom layer is critical. The maximal thermal budget for subsequent steps is limited by salicidation. In fact, standard NiSi dewetting occurs in only 3 min at 600°C. To obtain a low resistance access with processes around 600°C, a salicide stabilization procedure has been demonstrated in [45]. This salicidation approach is based on an optimized W-NiPtSi-F. With this technology, the salicide is stable at 600°C, which is considered to be the maximal thermal budget for all the subsequent steps.

The second step corresponds to the realization of a high quality top film. While seed window techniques were first used in the literature [46], they lead to crystalline defects, poor thickness control and density loss. It is thus better to used

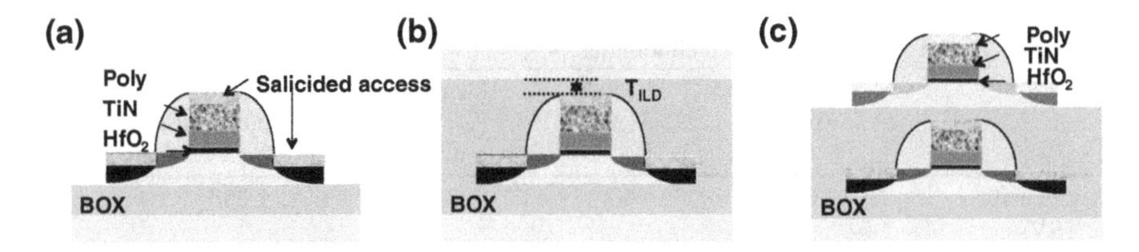

Fig. 3.11 Cross-sectional view of 3D monolithic steps—**a** optimized bottom FDSOI process **b** high quality top film deposition **c** low temperature top FDSOI process [44]

molecular bonding [45], where a blanket Silicon-On-Insulator wafer is transferred on top of the processed MOS wafer. This step plays a major role in alignment accuracy, since the alignment occurs after bonding, conversely to parallel integration [47].

A low temperature Fully Depleted Silicon-On-Insulator is then processed on the top film. Due to the limited thermal budget (600°C), it is not possible to perform thermal activation of dopants (around 1,000–1,100°C). The process uses a solid phase epitaxial re-growth [48], which consists of pre-amorphization of the top film, followed by dopant implantation and finishing by a re-crystallization at 600°C.

Finally, 3D contacts are realized using the same contact techniques as in standard processes. Only one lithography step is required for all contacts. The contact realization step is illustrated in [45], where W (tungsten) plugs exist between the top and bottom layers. The process has been validated for several simple cases, such as SRAM memories or inverter cells [48, 49].

3.3.3 Elementary Blocks

In order to evaluate the opportunity of the monolithic 3D integration process, elementary nodes for configurable logic have been designed. The blocks that have been chosen are a 3D Look-Up Table and a 3D Cross point. The global idea is to place the configuration memory just above the circuit that requires it.

3.3.3.1 LUT Impact

Our design assumption is that the top layer is used to implement configuration memories in a standard CMOS approach. Hence, it is quite straightforward to develop a two layer 3D look-up table. In Fig. 3.12, the layout and the schematic of a 1-bit Look-Up Table are depicted. While the structure is too simple for any architectural considerations, it is useful for the purposes of technological demonstration.

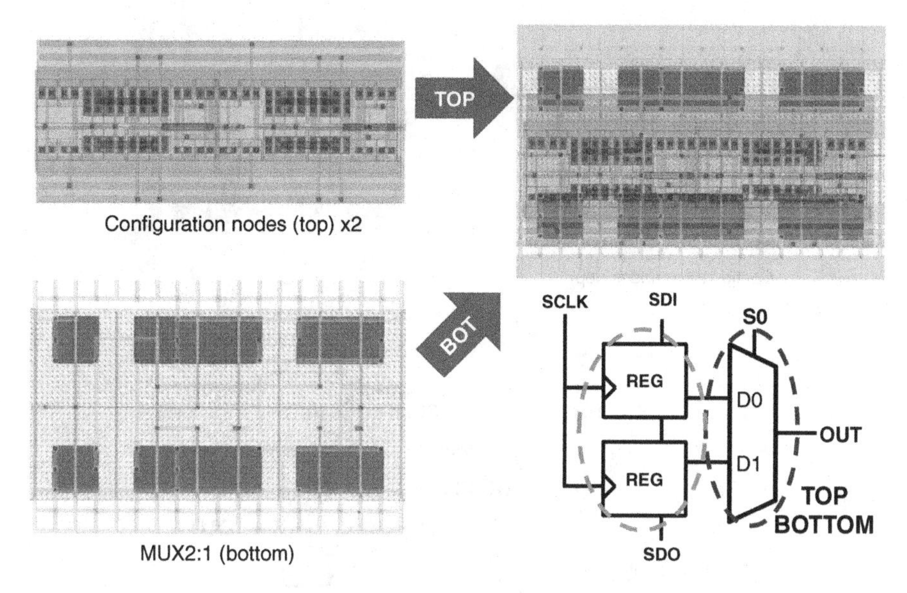

Fig. 3.12 Two layer monolithic 3D based 1bit Look-Up Table—layout and schematic

Two configuration nodes are thus placed above a multiplexer circuit. The dimensions of the multiplexer are largely relaxed, due to limitations on the metal layers available for the technological demonstrator.

In this way, we could consider that only a small amount of size is used for the structure, in addition to the multiplexer. Furthermore, while we observe a strict separation between the configuration memory part and the data path multiplexer, it is possible to optimize the technological process to reach specific properties. Especially, it is envisaged to have a low-leakage—low-power process for memories, while the active data paths are created using high performance transistors. This makes the implementation highly promising for circuits requiring large configuration memory.

3.3.3.2 Routing Structure Impact

Using the same technological assumptions, we also propose a routing cross point. The cross point is built around a two-transistor pass-gate, which is driven by a static memory. The memories and data path will be split and placed on different levels. Memories are placed on top of the cells that directly use their information, and thus a large reduction in the number of wires is observed. Two different structures have been realized.

Figure 3.13 shows the cross point using register-based memory. It is worth noticing that, in this circuit and conversely to the LUT, the memory dominates the dimensions of the cell. Indeed, the pass-gate, even when considering a large size transistor, has low requirements in terms of the metal layers, and thus its

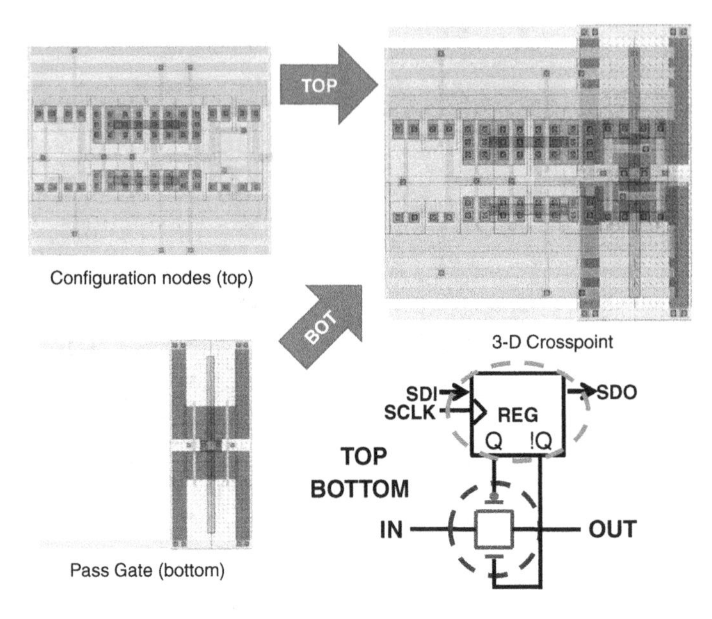

Fig. 3.13 Two layer configurable cross points (configuration node based)—layout and schematic

implementation below the memory is kept at the front-end. We also point out the H-shape of the pass-gate, which is a side effect of the optimization of contacts between the bottom, the top and the first metal layers.

The above implementation uses a configuration node, which could be programmed serially. This is obviously of high interest for FPGAs, since it allows memory cascading and thus simplifies the programming circuitry. Nevertheless, in the context of separation between logic and configuration memory in a 3D process, it is of interest to use a standalone memory approach. Hence, a typical SRAM organization could be envisaged on the top layer with word lines and bit lines crossing through the circuit, while the FPGA data path is found below. Figure 3.14 shows a possible implementation of an SRAM memory with a pass-gate in a very compact 3D implementation.

3.3.4 Performance Characterization

3.3.4.1 Methodology

The performance metrics of the presented elementary blocks have been extracted from layout and electrical simulations.

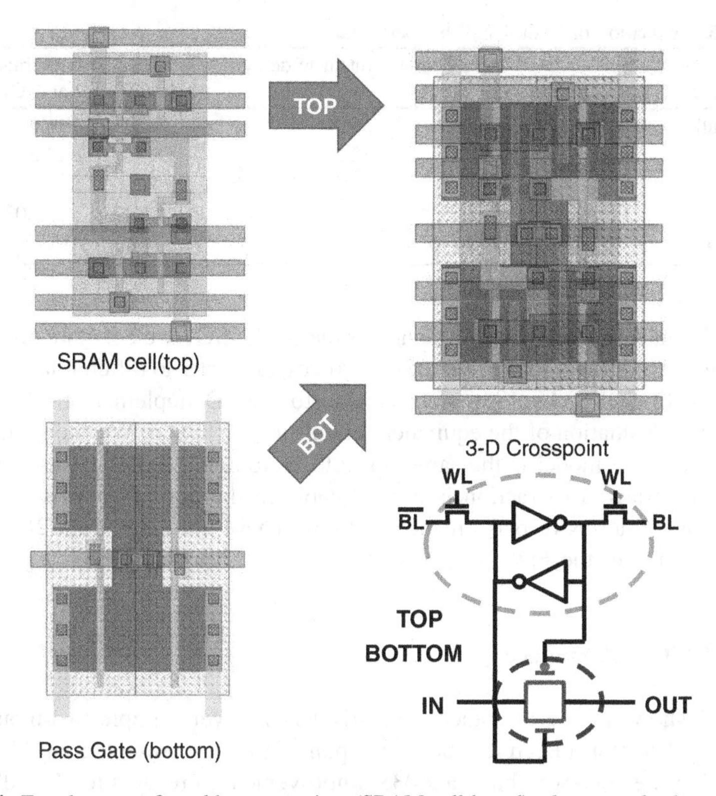

Fig. 3.14 Two layer configurable cross points (SRAM cell based)—layout and schematic

The simulations are realized using the Synopsis HSPICE electrical simulator. This simulator allows a fast and accurate simulation for CEA-LETI's homemade model.

The metrics will be area, delay and power consumption. The area is extracted considering the layouts. The test chip uses two different set of design rules for front-end and back-end. In order to demonstrate the capabilities of the technology with a relatively low cost, the front-end can be scaled down to the 16-nm node, while the back-end uses a 65-nm lithographic node. Nevertheless, in order to be compared fairly with existing technology, all the dimensions used in the layouts use an equivalent 65-nm lithographic node. The delay is assessed by electrical simulation and broken down into the intrinsic delay and the load delay (expressed as the K_{load} factor). The power consumption is extracted by electrical simulation with an FO4 load. The circuit is operated at the gate's maximum achievable frequency with the FO4 load. This means that rise and fall time are chosen in accordance to 5τ charge and discharge. The related frequency obviously depends on the circuit. At this frequency, we swept all possible input vector combinations and we averaged the power consumption.

Table 3.5 Evaluation of Look-Up Table performance

LUT2 65-nm node	Area (μm^2)	Intrinsic delay (ps)	K_{Load} (ps.fF^{-1})	Average power at 1 GHz (μW)
2D LP bulk	64.48	85.08	36.29	46.90
2D FDSOI	122.57	52.52	5.94	23.23
3D FDSOI	31.7	52.52	5.94	23.23
2D bulk versus 3D FDSOI	× 2.03	× 1.62	× 6.11	× 2.02
2D FDSOI versus 3D FDSOI	× 3.87	–	–	–

In order to compare the performance of the 3D FDSOI, we compare all metrics to a standard industrial low power 65-nm process. In order to differentiate the gain due to the FDSOI technology and that due to the 3D implementation, we also perform the evaluation of the equivalent circuit in 2D FDSOI. We should mention that the electrical model is the same for both bottom and top transistors and that parasitic post-layout extraction is not available in the design kit as used. Thus, comparisons in terms of performance and power will not differ from 2D and 3D implementation of the FDSOI cells.

3.3.4.2 LUT Performance

Table 3.5 shows the performance comparisons for a very simple two-input LUT test case. The considered circuit corresponds to that presented in Fig. 3.12. Concerning the area, we obtain a 2.03× improvement in regards to 2D bulk. This figure is mainly due to the stacked integration of the memory on top of the multiplexer. It is also worth mentioning that the 2D FDSOI implementation is larger than the 2D bulk one. This result is counter-intuitive, since the lithographic node used is the same in both cases. However, we should note that the 2D FDSOI layout has been carried out following regular layout techniques, as well as some relaxed technological demonstrator rules. This clearly leads to larger cell implementations than in an equivalent non-regular bulk process.

In terms of performance and power, an improvement of 1.62× in intrinsic delay can be observed, as can figures of 6.11× in load influence factor and 2× in dynamic power at 1 GHz. The contributing reasons for these good numbers are twofold. As already stated, the technology is compliant to low-power high-performance circuits. Nevertheless, the improvements are also due to 3D integration. It is possible to group the low-power optimized circuits on one layer, while performance-optimized blocks are grouped on another. This allows specific process optimization, but it also allows the dimensions of some transistors to be relaxed. For example, in this context of LUT, it is possible to size the multiplexer to be quite large, while the memories are placed on top. This strategy leads to a high-performance multiplexer, placed under the necessarily area-hungry memories.

Table 3.6 Evaluation of cross point performance

65-nm node	Area (μm^2)	Intrinsic delay (ps)	K_{Load} ($ps.fF^{-1}$)	Average power at 2 GHz (μW)
2D LP bulk	15.83	0.44	0.71	0.12
3D FDSOI Configuration node	10.66	0.28	0.67	0.039
3D FDSOI SRAM cell	5.40	0.14	0.66	0.058
2D bulk versus 3D FDSOI Conf. node	× 1.48	× 1.6	× 1.06	× 3.1
2D bulk versus 3D FDSOI SRAM cell	× 2.93	× 3.1	× 1.07	× 2.1

3.3.4.3 Cross Point Performance

Table 3.6 shows the performance evaluation of 3D cross points. As previously described, two different cross point schemes are studied. The difference between the two cross points resides in the configuration. One is based on a shifter configuration node and the other is based on an SRAM.

In terms of area, we should note that the SRAM-based cell is the most compact one, with an improvement factor of 2.93× compared to 2D bulk. This is obtained mainly due to the compactness of SRAM cells, where each transistor and layout is optimized to obtain the best density. Nonetheless, due to its greater size, the configuration-node based cross point improves the area by "only" 1.48×. It is worth noticing that this cross point implementation embeds all the circuitry required to program the node in a shift register manner. Thus, it appears obvious that this implementation is larger compared to the fully optimized SRAM-based implementation.

In terms of performance, the intrinsic delay is shortened by 1.6× and 3.1× for the configuration-node based implementation and the SRAM-based implementation, respectively. The load factor is almost the same with a gain of 1.06×. The difference comes mainly from the pass-gate cell. It is in fact possible to use larger transistors in the SRAM-based implementation; indeed, since the cell is more compact, the shape of its layout facilitates its placement and efficient connection of the bottom pass-gate.

Finally, in terms of dynamic power, we observe a 3.1× improvement at 2 GHz for the configuration-node based cell, and 2.1× for the SRAM-cell, compared to bulk. These numbers come mainly from the use of FDSOI, which is an intrinsic low power technology. The difference between the two memory structures is due to pass-gate sizing. Since most of the contribution to power originates in the data path, these results can again be attributed to the sizing of the pass-gate. In the SRAM-based case, the transistor is larger and leads to larger power consumption during the signal drive.

3.3.4.4 Comments

The described performance levels have shown a significant advantage of this technology for reconfigurable circuits. In general, the FDSOI technology is highly suitable for power reduction in electronic circuits. In particular, it improves the FET properties, which leads to reduction in leakage current, while the performance is increased. Nevertheless, the ability to stack several layers with a high alignment accuracy and high via density also allows to achieve high improvements in terms of area. In particular, we have shown a reduction in area by a factor of 2× for a 2-bit LUT test case circuit. The remaining area occupation is composed essentially of the bottom multiplexer. Furthermore, we can observe that the implemented circuits are following very strict (conservative) design rules, which are required by the technologist at the current level of technological maturity. In the future, it is clear that rules will be more aggressive, and will lead to further improvements in the figures.

3.4 On the Use of Vertical Silicon Nanowire FET Process

In the previous part of this chapter, we have investigated various technologies, which allow respectively a passive resistive element to be embedded in 3D and to stack several layers of active devices. Nevertheless, these solutions should not really be considered "true" 3D solutions since they cannot place active devices in a real 3D shape, i.e. with a vertical rather than planar transistor. In this section, we will assess such a technology, which is able to build vertical FETs in the Back-End metallic layers.

3.4.1 Introduction

As previously discussed, the 3D integration of transistors is an attractive solution to pursue the increase of circuit performance, while limiting the cost, as opposed to the continued single use of scaling. Stacking technologies, whether the traditional sequential technology (where wafers are processed separately and then stacked) or advanced monolithic integration (where transistors are processed step by step on the same wafer) only deal with stacks of planar transistors. The transistor itself (or more specifically, the transistor channel) does not exploit the vertical direction in these approaches.

Meanwhile, semiconducting nanowires have recently attracted considerable attention. To further miniaturize the transistor while still maintaining control over power consumption, alternative transistor geometries have been considered [50]. With their unique electrical and optical properties, they offer interesting perspectives for basic research as well as for technology. A variety of technical applications, such as nanowires as parts of sensors [51], and electronic [52, 53] and photonic devices [54] have already been demonstrated. In particular, electronic applications are increasingly coming into focus, as ongoing miniaturization in microelectronics demands new innovative solutions. Typically, silicon nanowire transistors have a horizontal, planar layout with either top or back gate geometry [55]. However, the amount of energy and time required to align and integrate these nanowire components into high-density planar circuits remains a significant hurdle for widespread application. More advanced works show a Gate-All-Around (GAA) organization in a planar topology [52]. In-place growth of vertically aligned nanowires, on the other hand, would in principle significantly reduce the processing and assembly costs of nanowire-based device fabrication, while opening up opportunities for "true" 3D. Some research works have demonstrated the possibility of fabricating transistors directly between two metal lines, within the back-end levels [56, 57]. These works make it possible to realize computation directly in the metal levels, through programmable vias, as well as potentially complete complementary logic functions.

3.4.2 Technological Assumptions

Recent studies have demonstrated the possibility to grow single crystalline silicon nanowires on a metallic line, into a CMOS compatible process [58]. This work represents a great opportunity to build FET devices in the interconnect levels [56, 57]. We propose to co-integrate standard CMOS with vertical Nanowires Field Effect Transistors. The cross-sectional view is shown in Fig. 3.15.

First of all, standard transistors are processed using the specified technology, which could be very versatile, such as Bulk, Silicon-on-Insulator, Fully Depleted SOI, Thin Box FDSOI, among others. Then, silicon nanowires can be grown in a CVD reactor using the VLS mechanism. Even on a metallic line, they have a single crystalline structure and semiconducting properties. Taking advantage of this, and respecting low temperature processes under 400°C, it is possible to make vertical transistors between two interconnecting lines. After etching a hole through the oxide to the metallic bottom line, a catalyst can be deposited at the bottom. Nanowires can be grown from the metallic line using the oxide hole as template and a deposited metallic catalyst. Using diborane or phosphine, nanowire can be doped to form P–N junctions. Nanowires for p-MOS and n-MOS should be grown during two distinct sequences comprising template formation and growth. After growth, a chemical etching can be used to remove a part of the oxide template. A multilayer gate stack can be achieved thanks to ALD and CVD deposit of the dielectric (Al_2O_3, HfO_2...), and the metal gate (TiN...) respectively. An oxide can then be deposited before performing a CMP step on the top. Isotropic etching allows the removal of a part of the metal gate and defining the gate length. The space left by the metal gate can be filled by oxide deposition. The top contact is achieved by top line formation using a conventional damascene process.

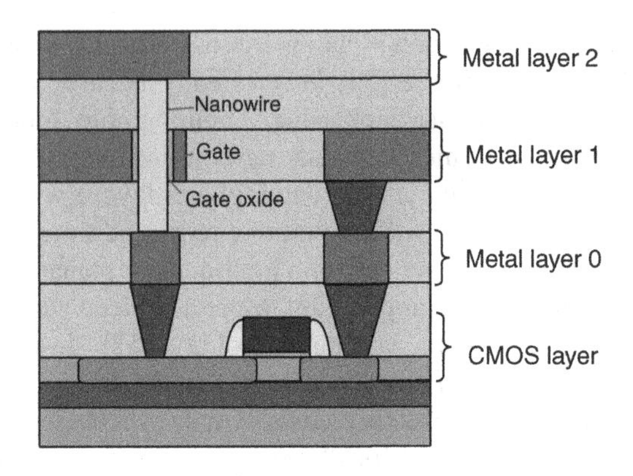

Fig. 3.15 Cross sectional schematic showing a BEOL FET and standard CMOS FET co-integration

3.4.3 Vertical 3D Logic

Since the described process could be used to integrate transistors into the metallic layers of a standard circuit, we can envisage adding some functionality, typically in relation to the back-end layers directly above the circuit.

3.4.3.1 Smart Vias

The association of vertical transistors and reconfigurable circuits leads immediately to the concept of "smart" reconfigurable vias. Here, the connection between a metal line and another is done by connecting the line to a transistor-drain and the other to a transistor-source. Thus, the connection is only controlled by the gate. This requires the line to be connected to the front-end and requires the use of several layers of vias to contact the back-end lines to the transistors. Here, based on the 3D FET technology, Fig. 3.16 shows an implementation of a controlled contact between two metal lines controlled by a combinational logic function.

In the presented illustration, four vertical wires are shown. Two transistors per wire are built in series. Four control gates drive the wires in a meshed pattern, which are used to create a combinational equation for the via conduction. The functionality is thus increased. The considered metal lines are connected when all transistors of any vertical branch are conducting. This corresponds to the following Boolean condition:

$$(G_A + G_B) \cdot (G_1 + G_2) = 1$$

3.4.3.2 Logic Gates

From the previously presented fully configurable via pattern, complementary logic cells can be built by including the dual branch using *p*-type transistors. Based on

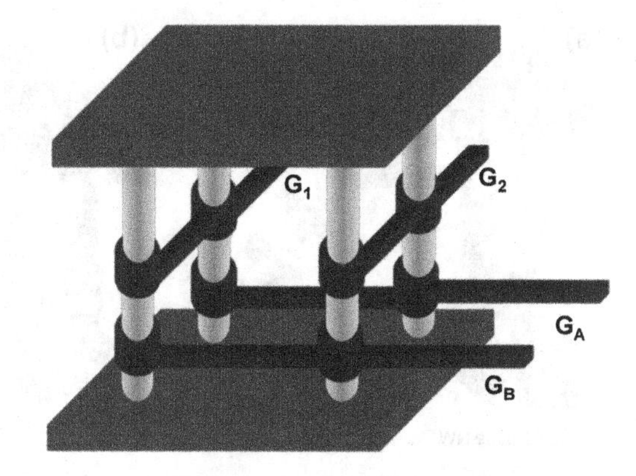

Fig. 3.16 Boolean logic with vertical FETs

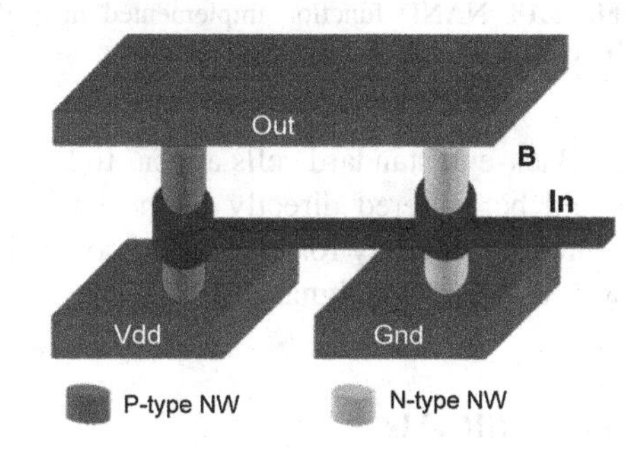

Fig. 3.17 NOT function implemented in a 3D standard cell

the 3D FET technology, we propose a fully back-end implementation of standard logic gates.

As an illustration, Fig. 3.17 shows a NOT gate built with vertical transistors. The circuit is formed of two vertical transistors of different doping types. An *n*-type transistor is built between the ground and the output lines, while a *p*-type transistor is placed between the power supply and the output lines. The peculiarity of the structure is that output lines and the power lines are separated by a metal layer. This metal layer is used to realize the transistor gates. It is worth noticing that different nanowire diameters could be used in the transistor channel, in order to size the structure according to design specifications. The proposed layout can be extended to any complementary logic gate.

We show in Fig. 3.18 the implementation of a NAND gate. Various actual layouts could be envisaged. Figure 3.18a shows an implementation where the serial transistors are shared on the same nanowire. While this solution is compact, it is worth mentioning that it also increases the vertical dimensions of the gate, because of the occupation of two levels of metal. Another solution is depicted in Fig. 3.18b. In this implementation, we only have one transistor per nanowire, i.e. between two metal lines. This allows the dimensions of the cell to be controlled, by analogy with the 2D standard cells, where only one level of metal is used for the interconnect.

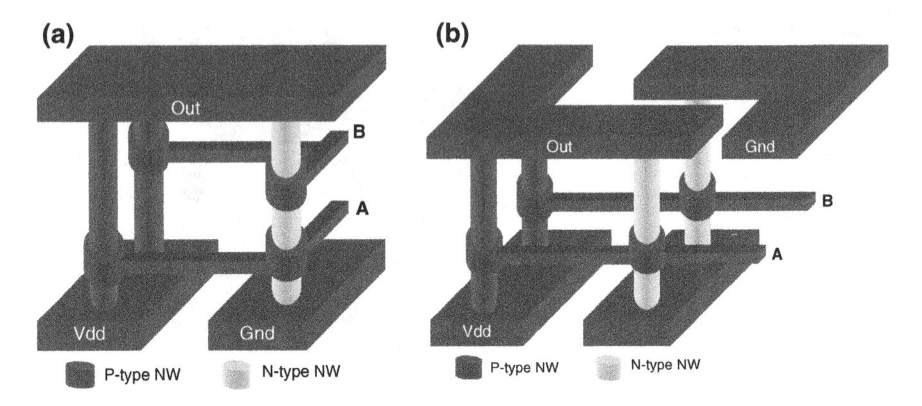

Fig. 3.18 NAND function implemented in a 3D standard cell **a** layout on two gate levels **b** layout with only one gate per nanowire

Back-end standard cells appear to be of use for routing application. Signal lines could be buffered directly in the back-end, thus avoiding connecting front-end buffer cells merely for routing purposes. This makes sense for long signal lines, as well as for clock signals.

3.4.3.3 SRAMs

Since we have shown that we could embed any logic function in a real 3D approach, it then appears quite straightforward to build memory elements. Figure 3.19 shows a possible organization for a 5T-based SRAM cell. Using the same approach as for the NAND gate, the layout is realized only between two metallic layers. The interest of these elements will be twofold. Firstly, they will be used as configuration memories for the reconfigurable logic front-end. As already mentioned, positioning memory resources above the circuit is of high interest for reconfigurable logic compactness and performance. Furthermore, it is also of use to have memory elements which are realized in the same technology as the smart vias presented above. These nodes will allow the configuration of the smart vias to be stored very close to their final use, and will thus lead to extremely compact cross point nodes entirely implemented in the back-end levels, as shown in the top view in Fig. 3.20. While the cross point uses a standard CMOS architecture as shown in Fig. 3.20a, its implementation is based on the use of smart vias and 3D SRAM cells. In fact, four 3D SRAM cells are used to store the configuration as close as possible to the logic circuit which uses the information. Two other SRAM cells implemented in 2D are also used. Thus, it is possible to mix the different technologies to obtain a structural trade-off between the requirements on the front-end area and on the routing ability. It is actually not advisable to embed all the logic in the back-end, but only to place a part of it in the most efficient way.

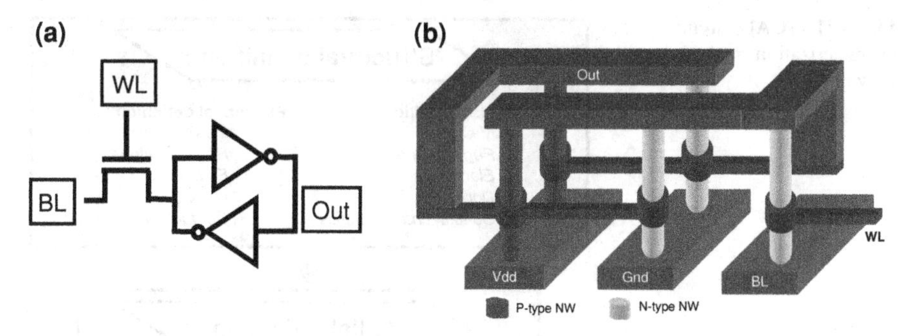

Fig. 3.19 5T SRAM cell implemented in a 3D standard cell— **a** schematic **b** cross view

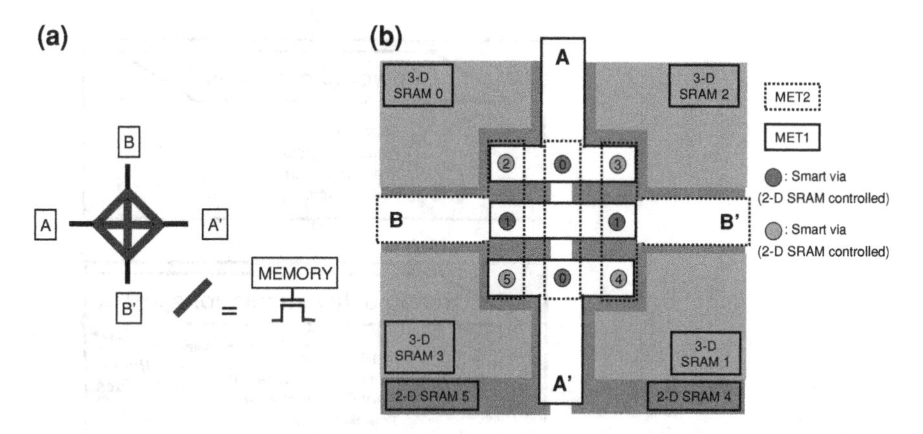

Fig. 3.20 Cross point implemented using 3D smart vias and 3D SRAMS **a** schematic **b** *top* view

3.4.4 Performance Characterization

3.4.4.1 Methodology

In order to evaluate the impact of the back-end implementation of standard cells, we investigate the area, delay and power consumption of elementary gates. Then, we compare our logic gates to the equivalent cells taken from an industrial 65-nm CMOS bulk process design kit.

Since the considered technology is not mature enough, there are no compact models available. The evaluation of 3D-BE standard cells have been performed thanks to TCAD ATLAS [59] simulations. TCAD simulations are computationally expensive, but they are flexible and allow new fundamental devices to be modeled efficiently. These characteristics mean that while they are widely used for physics and device simulation, they are rarely used for circuit design. Nevertheless, they represent new opportunities for design methodologies working close to the

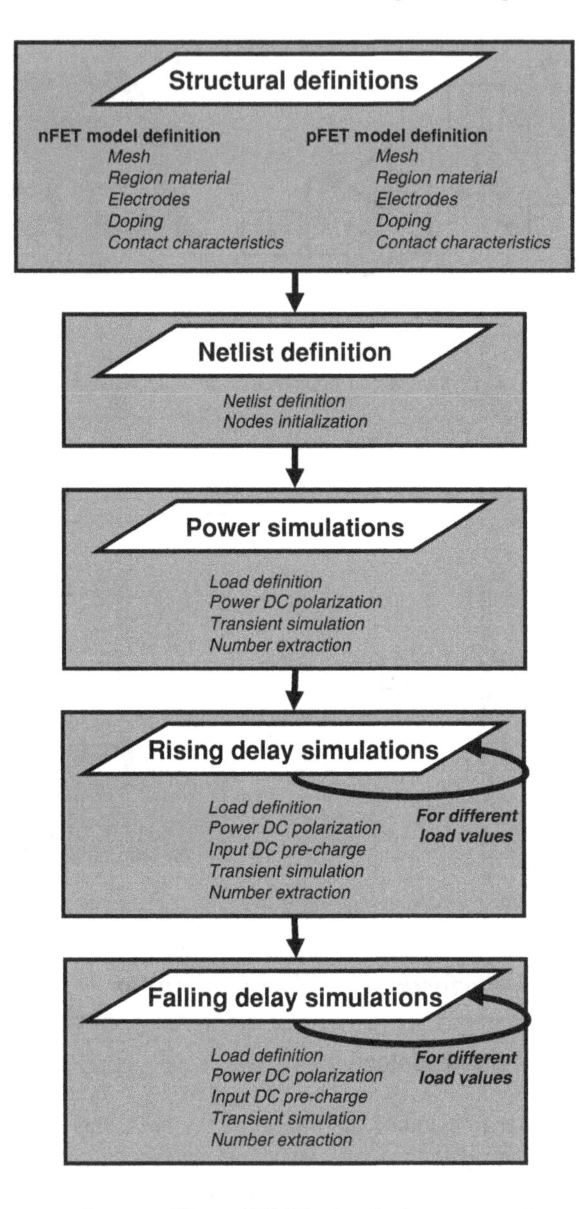

Fig. 3.21 TCAD circuit characterization methodology flow chart

technology. In our original approach, we will use TCAD simulations, not only to simulate the performances of the unique device, but also to simulate standard circuit and to extract its performance numbers.

Vertical nanowire transistor structures have been modeled according to the proposed device architecture. The circuit netlists are then input to the TCAD simulator and transient simulations are performed. From the transient simulation results, the performances metrics are then evaluated. The complete methodology is described in Fig. 3.21.

3.4.4.2 Device Characterization

In order to evaluate the technology at the lowest level, technological simulations have been used to extract the I–V curves of the elementary vertical transistors. Figure 3.22 depicts the I–V curves of an *n*-type and a *p*-type vertical FET. These curves are extracted for a NWFET with a 50-nm diameter, a 300-nm gate length and 5-nm oxide thickness. It is worth noticing that the devices present a very good I_{on}/I_{off} ratio in the range of 10^8 and a low I_{off} current less than 0.1pA. These excellent properties can be explained by the structure of the transistor. First of all, the transistor is a Gate-All-Around structure, which means that the active zone must be considered all around the nanowire. This makes the device highly controllable, while the dimensions are maintained very compact. The structurally optimal electrostatic control of the device leads to the low I_{off}. Furthermore, the structure uses the vertical dimension to increase drastically the dimensions of the active region, while the front-end impact is maintained very small. Thus, it is possible to realize large FETs, with excellent electrical properties and very compact front-end projection.

The estimated performance levels of the elementary transistor lead to the realization of high performance switches. For routing applications, it is of high interest to achieve pass transistors which exhibit a low I_{off} current, while the I_{on} current is in the range of several μA. It is clear that the higher the I_{on} current, the lower the R_{on} resistance and consequent switch impact on the overall circuit performance.

3.4.4.3 NOT Gate Characterization

While it is quite standard in TCAD simulation to extract the I–V curve of a device from the simulation, it is also of high interest to evaluate, from a transient simulation point of view, the characteristics of simple circuits. Table 3.7 summarizes the performance results of a NOT gate and compares it to a standard bulk cell. Such a comparison is helpful to evaluate the properties of the technology not only for the device but also for its ability to improve circuit performance.

We see that the proposed 3D implementation of a NOT gate clearly improves the area by a factor of 31.2×. These results come from the use of the vertical direction to implement the transistors. Delay and leakage power are also improved by a factor of 2.5× and 14.5× respectively. This can be explained by the good performance levels of the gate-all-around control of transistors. Indeed, the good electrostatic control of the channel helps to have a well defined off state, and thus to obtain a very low I_{off} current, while the I_{on} current remains high (the I_{on}/I_{off} ratio is high). Thus, since these characteristics are directly involved in the power consumption and delay estimations, the figures show clear improvement over the equivalent MOS technology.

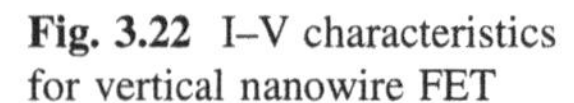

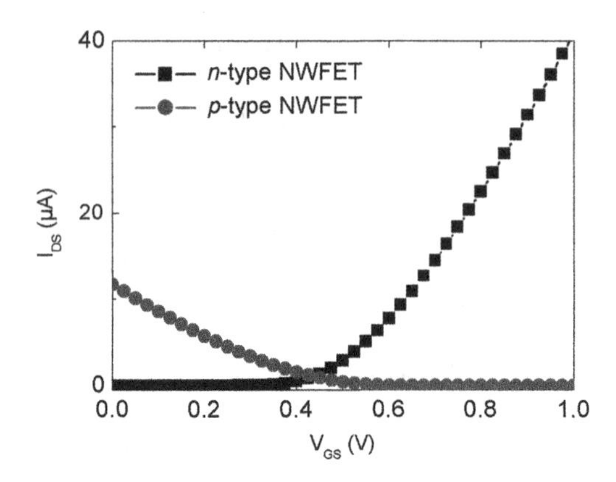

Fig. 3.22 I–V characteristics for vertical nanowire FET

3.4.4.4 Discussion

These characteristics show some promising performances for the technology. We have seen that the elementary device can be sized efficiently, thanks to the use of the third dimension. It is thus possible to obtain high performance devices, while the impacted area on the front-end is maintained very low. This opens the way towards high performance switching between two metal lines, which is of high interest for routing resources in reconfigurable circuits such as FPGAs. Furthermore, we have seen that it is possible to implement logic functions, using complementary logic with this technology, and that the obtained circuit performance gain is also significant with respect to CMOS. While the area is improved again by the use of the third dimension, we also obtain high performance figures, due to good device performance levels. It is thus of great interest to examine the way such logic could embed elementary active functions, as required for the routing circuits and clocks. For example, it is possible to embed the signal buffers as well as high performance clock tree buffers along the line which requires this functionality. This avoids inefficient multiple vertical communication schemes such as those seen today between the active transistor front-end and low-RC metal back end.

3.5 Global Comparisons and Discussions

Several technologies have been investigated in this chapter in order to move a part of the FPGA functionality into the back-end levels. All these techniques integrate an above–IC device. While they are all promising for routing resource improvement, it is interesting to compare them using the same template. To do so, we use a four-input LUT, which is the elementary block used in modern FPGAs to perform the computation.

Table 3.8 compares the performance of the structure realized with the various technologies. The assumptions for the realization differ from the technology, since

Table 3.7 Simulation results summary

	Area (μm^2)	Intrinsic delay (ps)	K_{Load} ($ps.fF^{-1}$)	Leakage power (pW)
NOT (MOS 65 nm)	1.6	17.5	3.8	40.1
NOT (3D BE)	0.05	6.9	4.7	2.8
3D BE versus CMOS	× 31.2	× 2.5	× 0.8	× 14.5

we need to consider the programming circuitry for a fair comparison. The bulk implementation serves as a reference. The storage elements are implemented as SRAM cascaded in shift registers, while the data path is composed of multiplexers. The ReRAM implementation uses multiplexers for the data path, but their data inputs are driven by the configuration nodes presented in zero. The programming structure has been incorporated into the evaluation, at least to ensure the fair selection between the power lines. The monolithic 3D FDSOI implementation uses multiplexers realized on the bottom silicon layer, while the SRAM shifted configuration memories are built on top. Finally, the vertical NWFET implementation uses standard 2D multiplexers to realize the data path, while the configurations are placed above the data paths using the circuit proposed in zero.

In terms of area, the most compact solution is obtained for the vertical NWFET technology. Indeed, the improvement of 10.2× is the consequence of the full vertical integration of the memory circuits. While the circuit is placed in the third dimension, its volume remains constant while the front-end projection (i.e. the area occupies by the drawn layout) is very low. In second place comes the ReRAM implementation with a gain of 7.3×. In this implementation, all the configuration memories are placed above the circuit, but programming access transistors are still in front-end silicon. Finally, the 3D FDSOI improves the structure by 5.7×, thanks to the two-stack repartition of transistors.

In terms of performance, the best figures are obtained with 3D FDSOI. In this case, the data path technology is moved towards enhanced silicon FDSOI. This leads to an improvement in the intrinsic delay of a factor of 4.4×, and of the load factor by 1.9×. For the other technologies, the performance levels are of the same order of magnitude as CMOS. This is due to the fact that the data path remains based on CMOS multiplexers and is thus the same between the different implementations. Nevertheless, it is worth noticing that the intrinsic delay is improved by 1.5×. This is because the data paths differ by one gate. Thus, the propagation delay is reduced by one intrinsic delay.

Finally, in terms of power consumption, only trends have been extracted, since an accurate estimation would require several simulations that are not possible to the unavailability of compact models for the vertical NWFET. Based on the previous results, we can remark that the ReRAM solution will remain of the same order of magnitude as the bulk solution. Indeed, while the data paths remain the same, the leaky SRAM circuits have been replaced by other potentially leaky memory nodes, such that the figures can be expected to be constant. Concerning

Table 3.8 Technology comparison (four-input LUT test case)

	Area (μm^2)	Intrinsic delay (ps)	K_{Load} ($ps.fF^{-1}$)	Average power gain trend
Bulk MOS 65 nm	547	465.6	11.2	Reference
ReRAM	74.5	310.4	11.2	=
3D FDSOI	95.1	105	5.94	× 2
Vertical NWFET	53.8	310.4	11.2	× 10
ReRAM versus CMOS	× 7.3	× 1.5	× 1	–
3D FDSOI versus CMOS	× 5.7	× 4.4	× 1.9	–
Vertical FET versus CMOS	× 10.2	× 1.5	× 1	–

3D FDSOI, power might be reduced by at least two, thanks to the low power characteristics of FDSOI. Finally, we have seen that vertical NWFET are large and possess good electrical properties. This provides them with a strong suitability for low power, and thus the trend can be estimated to be in the range of 10×.

3.6 Conclusion

In this chapter, we have investigated how disruptive technologies can be used to build enhanced basic logic circuits for routing and configuration. These elements are fundamental for FPGA technology since they represent the largest amount of required area (over 80%) and constitute the bottleneck for performance improvement.

Globally, the improvement of these circuits is based on the use of 3D integration technologies. In fact, we studied three different technologies, which respectively allow (i) passive reconfigurable elements to be placed in the back-end layers (ii) several layers of active silicon to be stacked with a high alignment accuracy and a high via density and finally (iii) "true" 3D transistors to be embedded in the metal layer in a vertical arrangement.

Resistive memories, and especially phase-change memories, have been initially considered. This technology is able to place a passive non-volatile resistive material in the back-end layer. This node can be reprogrammed between two stable resistive states. We propose a simple logic circuit based on two resistive nodes and one transistor in order to create a configuration node for reconfigurable logic. This node leads to an improvement of 1.5× in terms of area compared to its flash counterpart. Furthermore, we notice that the on-resistance of a resistive memory is very low compared to a CMOS transistor. We thus propose a switchbox structure that places the resistance directly in the logic data path. The solution is compact compared to the flash equivalent circuit with a gain of 3.4× in area, and we expect further improvements at the system level due to the reduction in the path resistance.

A monolithic 3D FDSOI integration process was then considered to embed active devices into the back-end layers. Such a technology is of particular interest for separating memory resources from the logic data paths in a reconfigurable device. In this context, it is possible to optimize the circuit technologies separately towards low-power or high-performance operation. Monolithic integration leads to a high via density, which allows fine grain architecture partitioning over the layers. To demonstrate the gain of the monolithic 3D integration, simple test structures were designed. We demonstrated a simple LUT structure, which improves the area by 2×, the delay by 1.6× and the power by 2× compared to its 2D CMOS counterpart. We also implemented a 3D cross point using two different configuration circuits. We demonstrated that this circuit yields an improvement of 1.5× in area, 1.6× in performance and 3.1× in power. This shows the benefits of the FDSOI technology when coupled to monolithic 3D integration, when compared to standard CMOS bulk.

Vertical NWFET has been finally assessed. This technology could be seen as the ultimate evolution to 3D, since it allows the integration of vertically oriented transistors. Thus, the designer can go beyond a stacked design of 2D devices, and can distribute the active devices within the back end layer with a very small impact on the front-end. This potential leads to the development of a smart back-end and the generalization of 3D computing. In order to evaluate the interest of the technology, we performed a TCAD evaluation for a simple circuit, and showed an improvement of 31× over CMOS in terms of area, while the delay is improved by 2.5× and the power by 14×.

To conclude this chapter, these technologies appear promising for reconfigurable circuit applications. Nevertheless, this kind of evaluation must be conducted fairly, thanks to the use of generic benchmarking tools and a set of well-known-application circuits. This evaluation will be the focus of the next chapter.

References

1. Xilinx Virtex-6 FPGA family overview, http://www.xilinx.com/support/documentation/data_sheets/ds150.pdf. 24 March 2011
2. J. McCollum, H.-S. Chen, F. Hawley, Non-volatile programmable memory cell for programmable logic array, US Patent No. 0,064,484, 2007
3. Emerging Research Devices and Materials Chapters, Updated Editions, International Technology Roadmap for Semiconductors (2010). http://www.itrs.net/Links/2010ITRS/Home2010.htm
4. H.J. Hovel, J.J. Urgell, Switching and memory characteristics of ZnSe–Ge heterojunctions. J. Appl. Phys. **42**, 5076 (1971)
5. I.G. Baek, M.S. Lee, S. Seo, M.J. Lee, D.H. Seo, D.-S. Suh, J.C. Park, S.O. Park, H.S. Kim, I.K. Yoo, U.-I. Chung, J.T. Moon, Highly scalable nonvolatile resistive memory using simple binary oxide driven by asymmetric unipolar voltage pulses, in *International Electron Devices Meeting*. pp. 587–590, 13–15 Dec 2004
6. K. Tsunoda, K. Kinoshita, H. Noshiro, Y. Yamazaki, T. Iizuka, Y. Ito, A. Takahashi, A. Okano, Y. Sato, T. Fukano, M. Aoki, Y. Sugiyama, Low power and high speed switching

of Ti-doped NiO ReRAM under the unipolar voltage source of less than 3 V, in *International Electron Devices Meeting*. pp. 767–770, 10–12 Dec 2007
7. C. Nauenheim, C. Kugeler, S. Trellenkamp, A. Rudiger, and R. Waser, Phenomenological considerations of resistively switching TiO_2 in nano crossbar arrays, in *10th International Conference on ULIS*. pp. 135–138, 18–20 March 2009
8. H.Y. Lee, Y.S. Chen, P. S. Chen, T. Y. Wu, F. Chen, C.C. Wang, P.J. Tzeng, M.-J. Tsai, C. Lien, Low-power and nanosecond switching in robust hafnium oxide resistive memory with a thin Ti cap. IEEE Electron Device Lett. **31**(1), 44–46, January (2010)
9. W.C. Chien, Y.C. Chen, K.P. Chang, E.K. Lai, Y.D. Yao, P. Lin, J. Gong, S.C. Tsai, S.H. Hsieh, C.F. Chen, K.Y. Hsieh, R. Liu, C.-Y. Lu, Multi-level operation of fully CMOS compatible WOX resistive random access memory (RRAM), *International Memory Workshop*. pp. 1-2, 10–14 May 2009
10. P. Zhou, H.J. Wan, Y.L. Song, M. Yin, H.B. Lu, Y.Y. Lin, S. Song, R. Huang, J.G. Wu, M.H. Chi, A systematic investigation of TiN/CuxO/Cu RRAM with long retention and excellent thermal stability, in *International Memory Workshop*. pp. 1–2, 10–14 May 2009
11. Z. Wei, Y. Kanzawa, K. Arita, Y. Katoh, K. Kawai, S. Muraoka, S. Mitani, S. Fujii, K. Katayama, M. Iijima, T. Mikawa, T. Ninomiya, R. Miyanaga, Y. Kawashima, K. Tsuji, A. Himeno, T. Okada, R. Azuma, K. Shimakawa, H. Sugaya, T. Takagi, R. Yasuhara, K. Horiba, H. Kumigashira, M. Oshima, Highly reliable TaOx ReRAM and direct evidence of redox reaction mechanism, in *International electron devices meeting*. pp. 1–4, 15–17 Dec 2008
12. A. Fantini, L. Perniola, M. Armand, J.-F. Nodin, V. Sousa, A. Persico, J. Cluzel, C. Jahan, S. Maitrejean, S. Lhostis, A. Roule, C. Dressler, G. Reimbold, B. De Salvo, P. Mazoyer, D. Bensahel, F. Boulanger, Comparative assessment of GST and GeTe materials for application to embedded phase-change memory devices, in *IEEE International Memory Workshop*. pp. 1–2, 10–14 May 2009
13. L. Perniola, V. Sousa, A. Fantini, E. Arbaoui, A. Bastard, M. Armand, A. Fargeix, C. Jahan, J.-F. Nodin, A. Persico, D. Blachier, A. Toffoli, S. Loubriat, E. Gourvest, G. Betti Beneventi, H. Feldis, S. Maitrejean, S. Lhostis, A. Roule, O. Cueto, G. Reimbold, L. Poupinet, T. Billon, B. De Salvo, D. Bensahel, P. Mazoyer, R. Annunziata, P. Zuliani, F. Boulanger, Electrical behavior of phase-change memory cells based on GeTe. IEEE Electron Device Lett. **31**(5), 488–490 (May 2010)
14. G. Betti Beneventi, E. Gourvest, A. Fantini, L. Perniola, V. Sousa, S. Maitrejean, J. C. Bastien, A. Bastard, A. Fargeix, B. Hyot, C. Jahan, J. F. Nodin, A. Persico, D. Blachier, A. Toffoli, S. Loubriat, A. Roule, S. Lhostis, H. Feldis, G. Reimbold, T. Billon, B. De Salvo, L. Larcher, P. Pavan, D. Bensahel, P. Mazoyer, R. Annunziata, F. Boulanger, On carbon doping to improve GeTe-based phase-change memory data retention at high temperature, *IEEE International Memory Workshop (IMW)*. pp. 1–4, 16–19 May 2010
15. S. Lai, Current status of the phase change memory and its future, in *IEDM Technical Digest*. pp. 225–228, Dec 2003
16. S. Raoux, G. W. Burr, M. J. Breitwisch, C. T. Rettner, Y.-C. Chen, R. M. Shelby, M. Salinga, D. Krebs, S.-H. Chen, H.-L. Lung, C. H. Lam, Phase-change random access memory: a scalable technology, IBM. J. Res. Dev. **52**(4-5), 465–479, (2008)
17. G. Bruns, P. Merkelbach, C. Schlockermann, M. Salinga, M. Wuttig, T. D. Happ, J. B. Philipp, M. Kund, Nanosecond switching in GeTe phase change memory cells, Appl. Phys. Lett. **95**(4), (2009)
18. G.Servalli, A 45 nm generation phase change memory technology, *IEDM Technical Digest*. pp. 113–116, 2009
19. J.H. Oh, J.H. Park, Y.S. Lim, H.S. Lim, Y.T. Oh, J.S. Kim, J.M. Shin, Y.J. Song, K.C. Ryoo, D.W. Lim, S.S. Park, J.I. Kim, J.H. Kim, J.Yu, F. Yeung, C.W. Jeong, J.H. Kong, D.H. Kang, G.H. Koh, G.T. Jeong, H.S. Jeong, K.Kinam, Full integration of highly manufacturable 512 Mb PRAM based on 90 nm Technology, in *IEDM Technical Digest*. p. 49–52, 2006

20. D. Ielmini, M. Boniardi, Common signature of many-body thermal excitation in structural relaxation and crystallization of chalcogenide glasses. Appl. Phys. Lett. **94**(09), 091906 (2009)
21. D. Ielmini, Y. Zhang, Analytical model for subthreshold conduction and threshold switching in chalcogenide-based memory devices. J. Appl. Phys. **102**(5), 054517 (2007)
22. G. Betti Beneventi, A. Calderoni, P. Fantini, L. Larcher, P. Pavan, Analytical model for low-frequency noise in amorphous chalcogenide-based phase-change memory devices. J. Appl. Phys. **106**(5), 054506 (2009)
23. T. Morikawa, K. Kurotsuchi, M. Kinoshita, N. Matsuzaki, Y. Matsui, Y. Fuiisaki, S. Hanzawa, A. Kotabe, M. Terao, H. Moriya, T. Iwasaki, M. Matsuoka, F. Nitta, M. Moniwa, T. Koga, N. Takaura, Doped in-Ge-Te phase change memory featuring stable operation and good data retention, in *IEEE International Electron Devices Meeting*. pp. 307–310, 10–12 Dec. 2007
24. B. Gleixner, F. Pellizzer, R. Bez, Reliability characterization of phase change Memory, in *10th Annual Non-volatile Memory Technology Symposium*. pp. 7–11, 25–28 Oct. 2009
25. T. H. Jeong, M. R. Kim, H. Seo, J. W. Park and C. Yeon, Crystal structure and microstructure of nitrogen-doped $Ge_2Sb_2Te_5$ Thin Film, Jpn. J Appl. Phys.**39**, 2775–2779, (2009)
26. Y. Lai, B. Qiao, J. Feng, Y. Ling, L. Lai, Y. Lin, T. Tang, B. Cai, B. Chen, "Nitrogen-doped $Ge_2Sb_2Te_5$ films for nonvolatile memory," J. Electron. Mater. **34**(2), 176–181, (2005)
27. A.L. Lacaita, D.J. Wouters, Phase-change memories, Phys. Status solidi (a), **205**(10), 2281–2297, Oct (2008)
28. A. Pirovano, F. Pellizzer, I. Tortorelli, A. Rigano, R. Harrigan, M. Magistretti, P. Petruzza, E. Varesi, A. Redaelli, D. Erbetta, T. Marangon, F. Bedeschi, R. Fackenthal, G. Atwood, R. Bez, Phase-change memory technology with self-aligned μtrench cell architecture for 90 nm node and beyond, Solid-State Electron. **52**(9), 1467–1472, Sept (2008)
29. K. SangBum, Z. Yuan, J.P. McVittie, H. Jagannathan, Y. Nishi, H.-S.P. Wong, Integrating phase-change memory cell with Ge nanowire diode for crosspoint memory—experimental demonstration and analysis, IEEE Trans. Electron. Dev. **55**(9), 2307–2313, Sept (2008)
30. S. Brown, R. Francis, J. Rose, Z. Vranesic, *Field-Programmable Gate Arrays with Embedded Memories*, (Kluwer Academic Publisher, 1992)
31. Executive Summary, Updated Edition, International Technology Roadmap for Semiconductors (2010). http://www.itrs.net/Links/2010ITRS/Home2010.htm
32. V. Betz, J. Rose, A. Marquart, *Architecture and CAD for Deep-Submicron FPGAs*, (Kluwer Academic Publishers, New York, 1999) p. 264
33. J.H. Kyung, N. Chan, K. Sungraen, L. Ben, V. Hecht, B. Cronquist, A novel flash-based FPGA technology with deep trench isolation, IEEE Non-volatile Semiconductor Memory Workshop. pp. 32–33, 2007
34. K. J. Han, N. Chan, S. Kim B. Leung. V. Hecht, and B. Cronquist, A novel flash-based FPGA technology with deep trench isolation, *IEEE Non-volatile Semiconductor Memory Workshop*. pp. 32–33, 26–30 Aug 2007
35. N. Bruchon, L. Torres, G. Sassatelli, G. Cambon, New nonvolatile FPGA concept using magnetic tunneling junction, in *IEEE Computer Society Annual Symposium on Emerging VLSI Technologies and Architectures*. pp. 6, 2–3 March 2006
36. J.Z. Sun, D.C. Ralph, Magnetorsistance and spin-transfer torque in magnetic tunnel junctions. J. Magn. Magn. Mater. **320**(7), 1227–1237 (2008)
37. Y. Guillemenet, L. Torres and G. Sassatelli, "Non-volatile run-time field-programmable gate arrays structures using thermally assisted switching magnetic random access memories, IET. Comput. Digit. Tec. **4**(3), 211–226, May (2010)
38. S. Onkaraiah, P.-E. Gaillardon, M. Reyboz, F. Clermidy, J.-M. Portal, M. Bocquet, C. Muller, Using OxRRAM memories for improving communications of reconfigurable FPGA architectures, in *IEEE/ACM International Symposium on Nanoscale Architectures (NanoArch)*. San Diego (CA), USA.08–09 June 2011
39. Y. Akasaka, T. Nishimura, Concept and basic technologies for 3D IC structure, in *International Electron Devices Meeting*. **32**, 1986, pp. 488– 491

40. A. Rahman, J. Trezza, B. New, S. Trimberger, Die stacking technology for terabit Chip-to-Chip communications, in *IEEE Custom Integrated Circuits Conference, CICC '06*. pp. 587–590, 10–13 Sept 2006
41. D. Henry, S. Cheramy, J. Charbonnier, P. Chausse, M. Neyret, C. Brunet-Manquat, S. Verrun, N. Sillon, L. Bonnot, X. Gagnard, E. Saugier, 3D integration technology for set-top box application, in *IEEE International Conference on 3D System Integration, 3DIC 2009*. pp. 1–7, 28–30, Sept 2009
42. P. Batude, M. Vinet, A. Pouydebasque, C. Le Royer, B. Previtali, C. Tabone, L. Clavelier, S. Michaud, A. Valentian, O. Thomas, O. Rozeau, P. Coudrain, C. Leyris, K. Romanjek, X. Garros, L. Sanchez, L. Baud, A. Roman, V. Carron, H. Grampeix, E. Augendre, A. Toffoli, F. Allain, P. Grosgeorges, V. Mazzochi, L. Tosti, F. Andrieu, J.-M. Hartmann, D. Lafond, S. Deleonibus, O. Faynot, GeOI and SOI 3D monolithic cell integrations for high density applications, in *2009 Symposium on VLSI Technology*. pp. 166–167, 16–18 June 2009
43. P. Batude, M. Vinet, A. Pouydebasque, L. Clavelier, C. LeRoyer, C. Tabone, B. Previtali, L. Sanchez, L. Baud, A. Roman, V. Carron, F. Nemouchi, S. Pocas, C. Comboroure, V. Mazzocchi, H. Grampeix, F. Aussenac, S. Deleonibus, Enabling 3D monolithic integration. ECS J. **6**, 47 (2008)
44. Y.-H. Son, J.-W. Lee; P. Kang, M.-G. Kang, J. B. Kim, S. H. Lee, Y.-P. Kim, I. S. Jung, B. C. Lee; S. Y. Choi; U-I. Chung, J. T. Moon, B.-I. Ryu, "Laser-induced Epitaxial Growth (LEG) Technology for High Density 3D Stacked Memory with High Productivity, in *IEEE symposium on VLSI technology*. pp. 80–81, 12–14 June 2007
45. S. E. Steen, D. LaTulipe, A.W. Topol, D.J. Frank, K. Belote, D. Posillico, Overlay as the key to drive wafer scale 3D integration, Microelectron. Eng. **84**(5–8), 1412–1415, May–August 2007
46. P. Batude, M. Vinet, A. Pouydebasque, C. Le Royer, B. Previtali, C. Tabone, J.-M. Hartmann, L. Sanchez, L. Baud, V. Carron, A. Toffoli, F. Allain, V. Mazzocchi, D. Lafond, N. Bouzaida, O. Thomas, O. Cueto, A. Amaral, S. Deleonibus, O. Faynot, Advances in 3D CMOS sequential integration, in *IEEE International Device Meeting*. 2009
47. M. Ieong, B. Doris, J. Kedzierski, K. Rim, M. Yang, Silicon device scaling to sub-10-nm regime, Science, **306**(5704), 2057–2060, (2004)
48. J. Hahm, C.M. Lieber, Direct ultrasensitive electrical detection of DNA and DNA sequence variations using nanowire nanosensors, Nano Lett.**4**, 51–54, (2004)
49. T. Ernst, E. Bernard, C. Dupre, A. Hubert, S. Becu, B. Guillaumot, O. Rozeau, O. Thomas, P. Coronel, J.-M. Hartmann, C. Vizioz, N. Vulliet, O. Faynot, T. Skotnicki, S. Deleonibus, 3D multichannels and stacked nanowires technologies for new design opportunities in nanoelectronics, *IEEE International Conference on Integrated Circuit Design and Technology and Tutorial,ICICDT 2008*. pp. 265–268, 2–4 June 2008
50. T. Ernst T. Ernst, L. Duraffourg, C. Dupré, E. Bernard, P. Andreucci, S. Bécu, E. Ollier, A. Hubert, C. Halté, J. Buckley, O. Thomas, G. Delapierre, S. Deleonibus, B. de Salvo, P. Robert, O. Faynot, Novel Si-based nanowire devices: will they serve ultimate MOSFETs scaling or ultimate hybrid integration?, in *IEEE International Electron Devices Meeting*. 2008
51. A.-L. Bavencove, G. Tourbot, E. Pougeoise, J. Garcia, P. Gilet, F. Levy, B. André, G. Feuillet, B. Gayral, B. Daudin, Le S. Dang, GaN-based nanowires: from nanometric-scale characterization to light emitting diodes, Physica Status Solidi (a), **207**(6), 1425–1427, (2010)
52. Y. Cui, Z. Zhong, D. Wang, W. U. Wang, C.M. Lieber, High performance silicon nanowire field effect transistors, Nano Lett.**3**, 149–152, (2003)
53. J. Goldberger, A. I. Hochbaum, R. Fan, P. Yang, Silicon vertically integrated nanowire field effect transistors, Nano Lett. **6**(5), 973–977, (2006)
54. V. Schmidt, H. Riel, S. Senz, S. Karg, W. Riess, U. Gösele, Realization of a silicon nanowire vertical surround-gate field-effect transistor. Small **2**(1), 85–88 (2006)
55. V.T. Renard, M. Jublot, P. Gergaud, P. Cherns, D. Rouchon, A. Chabli, V. Jousseaume, Catalyst preparation for CMOS-compatible silicon nanowire synthesis. Nat. Nanotechnol. **4**, 654–657 (2009)

56. ATLAS User's Manual, SILVACO, 2008
57. C. A. Moritz, T. Wang, Latching on the wire and pipelining in nanoscale designs, *3rd Workshop on Non-silicon Computation (NSC-3)*. June 2004
58. T. Wang, P. Narayanan, C. A. Moritz, Combining two-level logic families in grid-based nanoscale fabrics, in *IEEE/ACM International Symposium on Nanoscale Architectures (NANOARCH)*. October 2007
59. P. Vijayakumar, P. Narayanan, I. Koren, C. M. Krishna, C. A. Moritz, Impact of nanomanufacturing flow on systematic yield losses in nanoscale fabrics, in *IEEE/ACM International Symposium on Nanoscale Architectures (NANOARCH)*. June 2011

Chapter 4
Architectural Impact of 3D Configuration and Routing Schemes

Abstract In this chapter, the architectural impact of the 3D enhanced memories and routing resources were carefully studied. The traditional FPGA architecture was enhanced by the technologies presented in the previous chapter. The envisaged technologies move devices in 3D. Devices can be passive (e.g. resistive phase-change memories) or active (e.g. monolithic 3D integration or vertical NWFET). Performance estimations were carried out by benchmarking simulations of the improved FPGA architecture. The benchmarking tool is based on standard tools and tuned according to the technological parameters. We showed that, implemented in FPGAs, the resistive configuration memory node, coupled to the routing structure, yields a delay reduction up to 51%, thanks to the reduction of dimensions and low on-resistance of PCMs. This result was also reached by the vertical NWFET technology, because of the ability to size a large transistor vertically without a large impact on the projected area. In this case, the critical path delay may be reduced up to 49% compared to the traditional scaled MOS. Regarding the area metric, the best improvement was reached by the vertical NWFET technology with an improvement of about 46%. Vertical NWFET technology allowed moving all the peripheral circuits above the IC. By opposition, the PCM technology leads to a much tighter area improvement of 13%. Indeed, this technology requires a large programming transistor per node.Among the different technologies, we should remark that 3D monolithic integration process yields in an area improvement of 21% on average and in a delay improvement of 22% on average. Such a technology represents a good trade-off process for short term micro-electronics evolutions.

In the preceding chapter, we proposed several approaches to enhance the performance of routing and memory structures in standard fine-grain reconfigurable architectures, using 3D integration techniques, where part of the circuits is stacked above the conventional silicon layer. The use of resistive memories allows passive memory devices to be placed in the back-end layers. For active devices, we explored the use of monolithic 3D integration and vertical nanowire FETs, to

P.-E. Gaillardon et al., *Disruptive Logic Architectures and Technologies*,

DOI: 10.1007/978-1-4614-3058-2_4,

respectively stack 2D FDSOI devices in a 3D scheme, and orient the active device channels in the third dimension. We analyzed the improvement of routing and memory blocks in terms of footprint, write time, data path resistance, etc. While these alternative circuits have been shown of interest in terms of their intrinsic properties, it is also mandatory to examine how they improve performance in a complete environment. Hence, in this chapter, we will focus on real-life circuit benchmarking, using FPGA architectures enhanced with the proposed approaches. After describing the motivation for architectural evaluation, we will describe the benchmarking toolflow that allows the evaluation. Then, for all the previously introduced proposals, we present a specific organization and the results of the benchmarking.

4.1 Motivation and Global Methodology

Research into emerging technologies is generally focused on devices and simple circuits. Nevertheless, it is necessary to pursue the analysis and assess the amount by which the use of a new technology can improve a complete application test circuit. The cost actually required developing, stabilizing and ramping up a new process can be justified only if circuit performance improvement is significant. Estimating such performance gain at an early stage in technology development is thus a critical step to channel work on emerging technologies.

In this work, we investigate reconfigurable logic applications, which are well-suited to disruptive technology analyses, due to the high architectural regularity (mirrored in the regular arrangements of emerging technologies), and high flexibility, an important factor for fault-tolerant circuits in the context of unreliable devices.

In this chapter, we then propose an assessment of the impact of the elementary blocks designed earlier on the complete FPGA architectural scheme. Indeed, while the individual block performance metrics are improved, it is necessary to check their impact at the system level.

Thus, we will consider the conventional island-style FPGA scheme, according to its description in Chap. 2. In particular, the logic part of the architecture is formed by CLBs. The CLBs are built by ten BLEs of four-input LUTs. Twenty two inputs connect the CLB to routing lines. Hence, we will replace the MOS routing and memory elements by the proposed structures.

In order to evaluate the performance metrics, it is necessary to set up a benchmarking toolflow. Part of the task of benchmarking is to "program" the reconfigurable structure with a set of well-known circuits, which constitute a representative sample of most application tasks the circuit has to perform. The other part of the benchmarking task is to evaluate and compare several metrics for each circuit. The most commonly used metrics are: the area used for logic and for routing, and the critical path delay through the routing structures. Others can also be added, such as power consumption, scalability, and robustness to name a few.

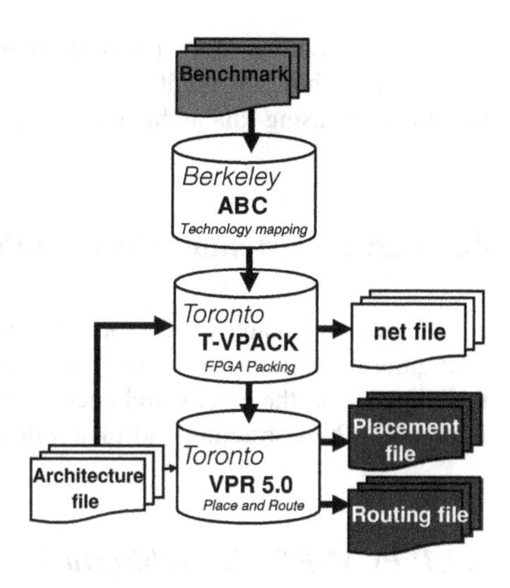

Fig. 4.1 FPGA benchmarking flow diagram

4.2 Benchmarking Tool for FPGA-Like Architectures

A conventional FPGA benchmarking flow is described in Fig. 4.1. The toolflow has been developed by the Toronto University [1], and is built around the T-VPACK and VPR5.0 tools. T-VPACK handles the packing of the logic blocks, while VPR5.0 handles the place and route of logic blocks in the island FPGA scheme. The input to T-VPACK is a netlist of LUTs and FFs. Thus, the logic packing operation consists of grouping the LUTs and FFs into BLEs, and then in packing several BLEs into CLBs. This operation should take into account all timing constraints. In particular, inter-CLB routing paths are faster than intra-CLB ones, such that it is possible to find an optimal packing arrangement that allows the connection delay to be globally improved. The netlist of CLBs, output from T-VPACK, is input into the VPR tool. VPR starts by placing the CLBs onto the island-style structure. The optimal position of the blocks over the grid is found using a simulated annealing algorithm [2]. Then, connections between CLBs are found using a pathfinder algorithm [2]. Obviously, all these steps use timing-based metrics. Finally, VPR outputs several metrics such as the total area used by the circuit and the critical path delay. To ensure the connection between the VPR flow and standard BLIF benchmarks [3], the ABC tool performs the logic optimization, the technology mapping onto a set of LUTs, and finally outputs the netlist of LUTs and FFs that is used by the flow.

This toolflow was initially developed for architectural exploration in the context of LUT-based FPGAs. It is therefore not possible to target an architecture that is not based on LUTs. Nevertheless, it is worth pointing out that the VPR tool is highly versatile in terms of architectural description, expressed in XML. The XML file consists of several sections that specifically describe part of the FPGA area. Hence, it is possible to specify the properties of the segment switches (i.e. the transistors that are used for routing), in great detail (albeit with a large number of parameters). Thus, the architecture can be adapted quite easily to advanced routing technologies, using this architecture fine tuning.

4.3 Resistive Memory-Based FPGA Performance

In the previous chapter, we proposed two circuits using resistive memories: the configuration memory node and the switchboxes. In this section, we will assess their impact on the FPGA architecture. First, we will draw a global view of the improved FPGA structure and then will discuss the benchmarking results.

4.3.1 PCM-FPGA Architecture

The potential improvement of the FPGA scheme is twofold. Firstly, concerning the configuration nodes, we propose to replace the area-hungry SRAM by a smaller and more efficient circuit. While this replacement is expected to reduce the size of the complete FPGA, it is also of high interest to take advantage of the low on-resistance of ReRAM. A switchbox implementation has been proposed, which uses resistive memories to replace the pass switches.

The PCM-based FPGA will use both of these proposals. Figure 4.2 gives an illustration of the structure. Configuration nodes will be used to feed the multiplexer inputs of each LUT. Furthermore, since such a configuration node is able to place a fixed logic level on a logic gate, it is also intended to be storage mechanism for routing multiplexers, which can be found at every level. A multiplexer is used in the BLE, in order to choose between the registered or the unregistered version of the LUT signal. Multiplexers are also used at the CLB level for the local interconnect. Finally, routing multiplexers are used in the connection boxes to select the signal to be connected to a CLB input. For all of these routing BLEs, the proposed configuration node will be used instead of an SRAM. Finally, all switchboxes will be replaced by the proposed PCM-based switchbox.

4.3.2 Methodology

To perform the evaluation, we used the standard FPGA benchmarking flow. A set of logic circuits taken from the MCNC benchmark [3] were first synthesized using the ABC tool [4]. We then performed the technology mapping with a library of 4-input LUTs ($K = 4$), also using ABC. We subsequently performed the logic packing of the mapped circuit into CLBs with $N = 10$ BLEs per CLB and $I = 22$ external inputs using T-VPACK [5]. Finally, the placement and routing were

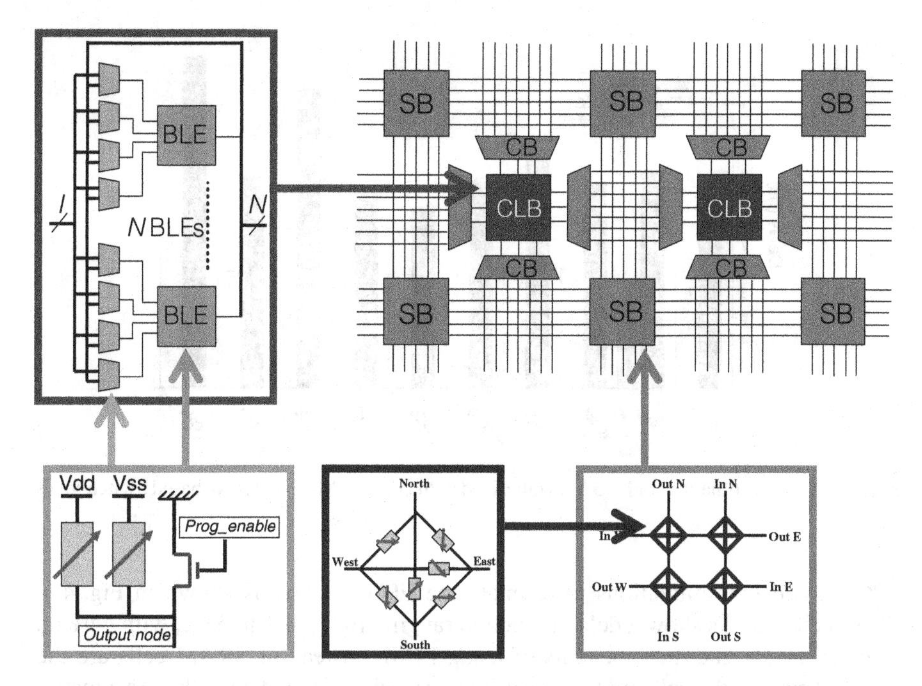

Fig. 4.2 PCM-based FPGA organization

carried out using VPR [5]. We synthesized the considered benchmark twice. The first (conventional) design was based on SRAM-based LUTs and MUXs in a 65-nm bulk CMOS process, using a pass-gate design. In the second (experimental) design, we replaced the SRAM cells in the LUTs and routing MUXs by PCMs. It is worth pointing out that, in routing topologies, PCMs are placed on the data path. Contrary to CMOS logic, which requires signal restoration, output buffers become superfluous with PCM technology.

4.3.3 Simulation of Large Circuits

In order to study the impact of each contribution, we analyzed two variants: one with an FPGA scheme improved by only the routing structures, and a second with a complete improved arrangement.

4.3.3.1 Impact on the Routing Structures

In this first variant, PCMs have been introduced only in the logic data path of the routing structures. We mapped the benchmarks onto both PCM- and SRAM-based

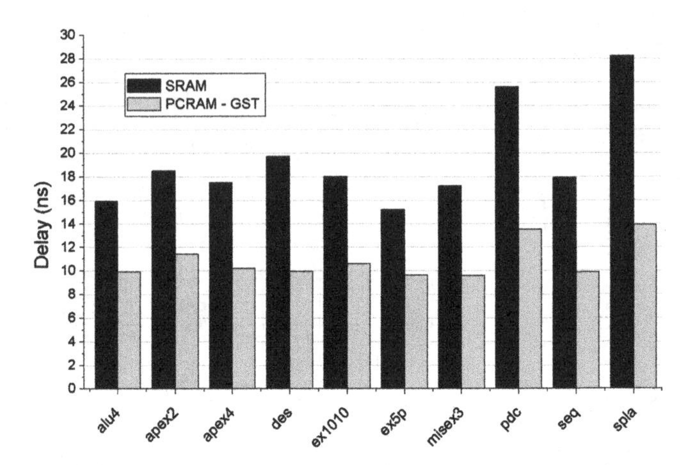

Fig. 4.3 Delay estimation for FPGAs synthesized with GST-PCM- and SRAM-based Switchboxes

FPGAs, and ran simulations to estimate the FPGA delay, as shown in Fig. 4.3. These benchmarks show a delay reduction ranging from 38% to 51%, with 44% on average. Hence the main benefits of using PCMs instead of SRAM cells are the compact area of the cell and the lower internal resistance of data paths. As a matter of fact, we extracted from our design kit the internal resistance of a pass-gate cell which is of the order of the on-resistance of an n-type transistor (9.1 kΩ); while the experimental results show that PCMs have a lower on-resistance that has been reported close to 4 kΩ [6]. This makes the PCM-based switchboxes potentially faster than the SRAM-based counterparts.

4.3.3.2 Impact on the Configuration Memories

We showed that the introduction of PCMs directly into the logic data path leads to an improvement in delay. In this section, we also add the memory node to the evaluation. Since routing is more compact than that of standard SRAM, we expect an area reduction. We mapped the benchmark onto PCM- and SRAM-based FPGAs, and ran simulations to estimate the FPGA area and delay. With respect to the previous evaluation, we have added more benchmarks in order to achieve a better test circuit diversity. We measured an area saving of about 13% on all our samples, due to the decrease in area allocated to memory. The delay estimation is depicted in Fig. 4.4, and shows a delay reduction ranging from 22 to 51%, with 40% on average. Here, the main benefits of using PCM-instead of CMOS-based cells are the compact area of the cell and its lower internal resistance. Again, we extracted from our design kit the internal resistance of a CMOS cell, of the order of the on-resistance of an n-type transistor (9.1 kΩ); while the measurements show that our PCMs have an on-resistance of 3.7 kΩ [6]. This makes the PCM-based

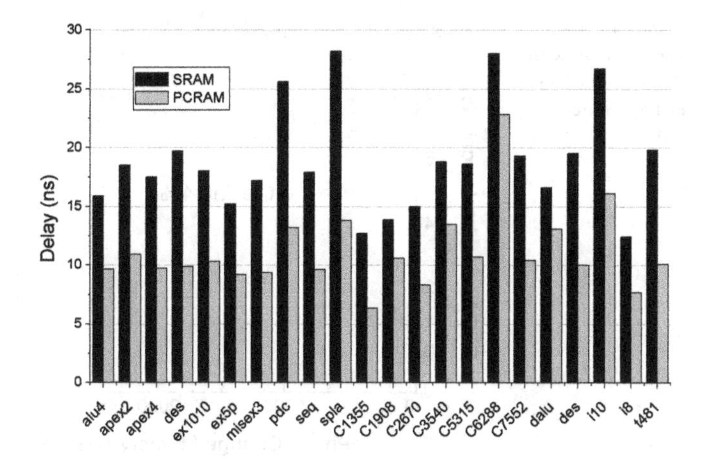

Fig. 4.4 Delay estimation for FPGAs synthesized with PCM- and SRAM-based LUTs and MUXs

FPGA potentially faster than the SRAM-based counterparts, given the lower resistive data path through the memory and the associated pass-gate MUX. Further, the compact area of the cell allows for a reduction of the CLB size and consequently a lower wire delay. The delay reduction due to smaller overall area is however less significant than the delay reduction due to faster routing elements.

4.3.4 Impact of Technologies

We demonstrated with the previous simulations that the combination of the switchbox design and the GST technology results in a significant delay improvement. In the following, we showcase the impact of the PCM technology type on the delay improvement. We remind the reader that other materials exist, which may replace the GST as a phase-change material. These include GeTe and GeTeCα%. Besides the difference in RESET current and time, the on-resistance depends on the chosen material and it has an impact on the routing path resistance. We simulated the delay of the FPGA benchmark with various resistance values corresponding to different materials. We notice that the delay improvement is linear with the decrease of the on-resistance. Figure 4.5 presents the simulation results. However, the delay sensitivity is low: a decrease in on-resistance by two orders of magnitude from about 4000 Ω to about 50 Ω causes a delay reduction of only 5%. The reduction of the on-resistance of the switches decreases the inter-CLB routing delay. In this situation, the intra-CLB routing delay becomes the dominant contribution and a larger reduction of the external switches on-resistance has no drastic impact.

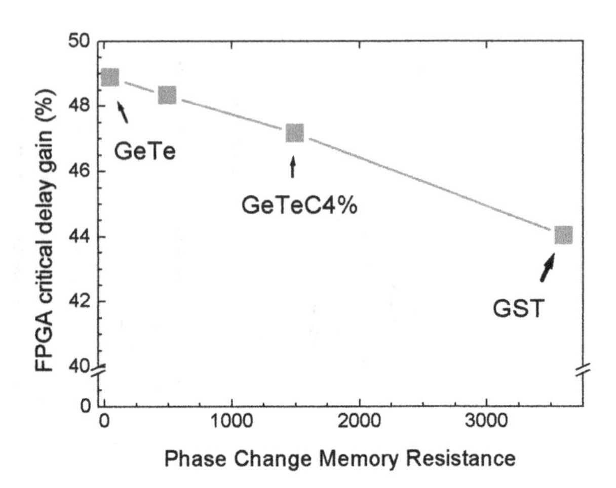

Fig. 4.5 Variation of the FPGA critical path delay with the PCM on-resistance (delay averaged over the whole benchmark set)

4.3.5 Discussion

In this section, we analyzed various performance metrics of a PCM-based FPGA. Our first observation is that the highest gain in terms of delay comes from the improvement of the routing structure. In fact, the reduction of the switch on-resistance yields an improvement of 40% on average. The configuration memory improvement is less significant and consists essentially of a reduction in terms of CLB area. This reduction appears quite low, at only 13% on average.

These results demonstrate significant benefits through the use of this technology. Indeed, it is worth pointing out that the technology is back-end of line compatible with CMOS technology. This makes it remarkable for its low-cost properties, while giving an improvement of 40% in delay.

We must also consider the limitation of the technology. Phase-Change memories require only a small amount of energy for their programming, which makes them power efficient. However, since the phase-change mechanism is based on the heat control of the material, the current density must be high. It is thus necessary to drive a large current into the memory node through a wide programming transistor, which remains costly in terms of silicon area. Furthermore, regarding the design of the configuration memory node, we should notice that there is inherent leakage, due to the off-resistance of the material, which must be kept as high as possible.

In order to retain the advantages of the modulated resistance, and mitigate the limitations, it is possible to consider other kind of ReRAM technologies. For example, it is worth noticing that Oxide-based memories could be programmed with a smaller current [7], while Conductive-bridge memories demonstrate a high R_{off} value [8].

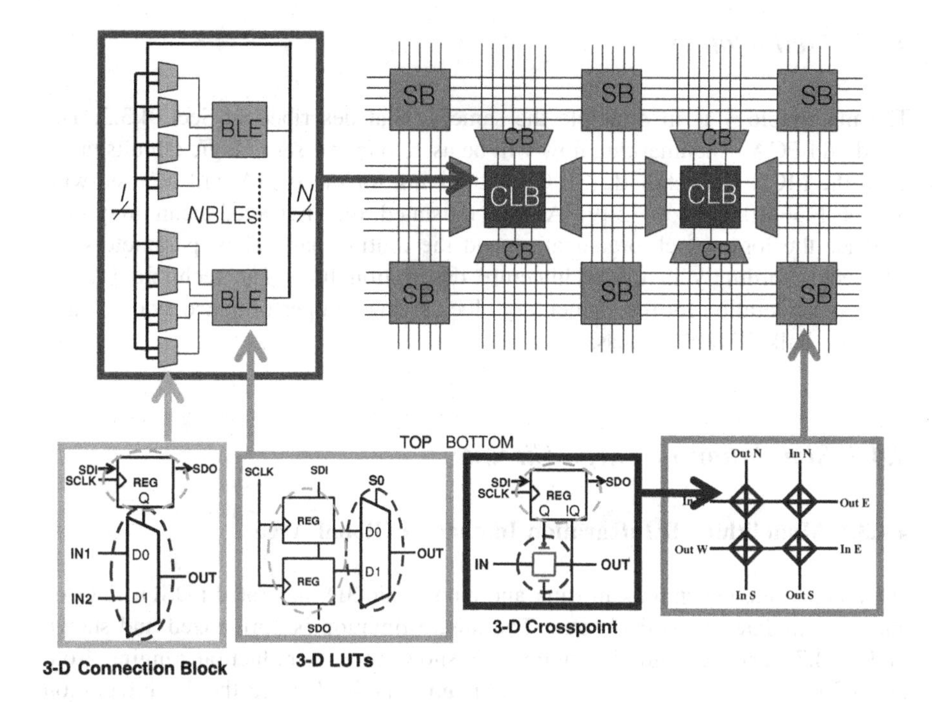

Fig. 4.6 Monolithically 3D integrated FPGA organization

4.4 3D Monolithic Integrated FPGA Performances

In the preceding chapter, monolithic 3D FDSOI basic blocks have been designed. In particular, memories were integrated on top of logic circuits. These blocks have been studied in comparison to standard FPGA blocks. In this section, we will assess the performance of a complete FPGA circuit, built in a 3D manner.

4.4.1 Overall FPGA Architectural View

In a CMOS FPGA, memories are found throughout the logic. Indeed, configuration memories are used to drive the LUT inputs, as well as to configure all the routing multiplexers. In the following FPGA evaluation, we will use the LUT scheme proposed in Sect. 3.3.3.2. As already stated, routing multiplexers can be found at each hierarchical level, from the BLEs to the Connection Boxes. Each routing multiplexer structure will use a top memory driving the configuration path of the multiplexer. All the switchboxes will use the scheme shown in Sect. 3.3.3.3. Figure 4.6 depicts a final view of the envisaged FPGA organization.

4.4.2 Methodology

The methodology is in principle the same as that described in Sect. 4.3.2. The standard FPGA benchmarking flow will be used to map a set of logic circuits taken from the MCNC benchmark [3]. Optimal sizing for the FPGA will be used with $K = 4$, $N = 10$ and $I = 22$ [9]. As already stated, the envisaged technology may enhance the logic block organization and the routing part. Thus, parameters are adapted accordingly in the architecture description files. The technology node under consideration for the elementary block sizing and performance evaluation is a 65-nm bulk CMOS process.

4.4.3 Simulation of Large Circuits

4.4.3.1 Monolithic 3D Integration Impact on Global Area

We mapped the benchmark in Bulk and monolithic 3D integrated FDSOI FPGAs, and we simulated the FPGA area. The area estimation is normalized and shown in Fig. 4.7. The benchmarks in Fig. 4.7 show an area reduction ranging from 20 to 25%, with 21% on average. The main benefit of using the 3D integration technology is the smaller impact on memory circuits on the area. Memories circuits are placed on top of the FPGA structure, while all the data path circuits are placed above. Nevertheless, since the top transistors are directly stacked above the bottom silicon layer, the circuit suffers some loss of performance in terms of routing. Indeed, the top transistors are impeding the connections between bottom transistors and the first metal layer. To improve the quality of routing and find the optimum routing organization, it is of interest to study the best trade-off in terms of internal metal layers.

4.4.3.2 Monolithic 3D Integration Impact on Critical Path Delay

The delay estimation with respect to the benchmark circuits is shown in Fig. 4.8. We can observe a delay reduction ranging from 10 to 45%, with 22% on average.

The main benefits of using the 3D FDSOI instead of standard Bulk are twofold. First, the intrinsic performance of FDSOI compared to bulk contributes to speed up the gate delay and thus reduce the computation time. Second, the reduction of routing area, as well as logic area, leads naturally to a reduction of routing wires. Shorter routing wires will be driven more efficiently, besides the fact that they will be controlled by drivers with improved output conductance. This makes the 3D integrated FPGA potentially faster than its standard Bulk counterparts.

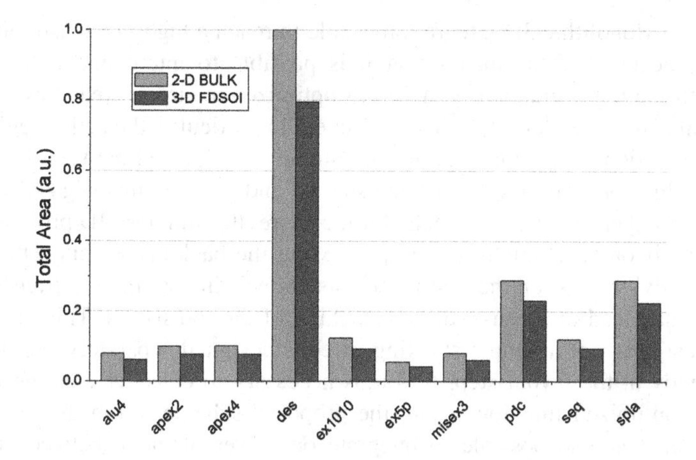

Fig. 4.7 Area estimation for FPGAs synthesized with standard bulk circuits and monolithic 3D integrated FDSOI circuits

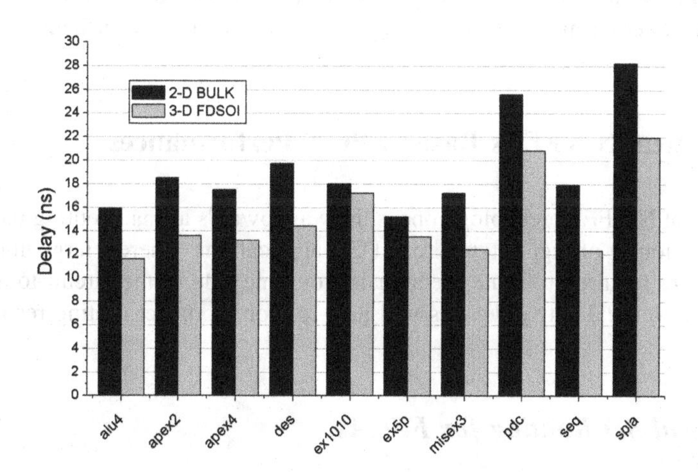

Fig. 4.8 Delay estimation for FPGAs synthesized with standard bulk circuits and monolithic 3D integrated FDSOI circuits

4.4.4 Discussion

Monolithic 3D integration appears to be an interesting technology from two aspects. First, it is currently reaching industrial maturity, which makes suitable for consideration for upcoming generations of reconfigurable circuits. In particular, it allows reductions in both the area of the global circuits and the critical path delay. The intrinsic performance of the FDSOI process is able to improve the speed of logic blocks, as well as the strength of the driver lines for routing resources. Furthermore, the integration on two layers clearly shows a reduction in the dimensions. Monolithic 3D integration is able to reach a high via density and high alignment accuracy. This means that it is possible to interconnect the various layers with the finest granularity, and thus optimize the circuit area. Nevertheless, the bottom and top layers are highly correlated. In particular, the active regions are very close, which means that significant coupling is expected between the layers. However, this coupling has never been studied and must be investigated in terms with silicon experiments and models. Furthermore, the simplest 3D process stacks the active silicon sequentially before processing the back end layers. This means that connectivity between the bottom transistors will impact the top transistors as well, and may lead to a non-optimum stacking of the transistors. This means that the 3D design is not a simple stacking of cells designed separately, but must be done directly at the layout step. It is thus necessary to consider creating real 3D cells, and optimized functions using the proposed scheme. From the technology perspective, it is also possible to integrate one layer of metal between the two active layers, and in the future, it will also be possible to integrate several layers between the active regions. This will improve the layout capabilities and lead to a smaller footprint. Further investigations will be required to find the optimal number of internal metal layers required for an efficient layout scheme.

4.5 Vertical NWFETs-Based FPGA Performances

The vertical NWFET technology opens the way towards a smart routing back-end. This is obviously of high interest for FPGA applications, where a large number of switches are required to route the signals from one side of the circuit to another. The use of vertical active devices will lead to more compact routing resources.

4.5.1 Real 3D Routing for FPGAs

The technology is intended to improve the FPGA routing resources, by pushing them into the third dimension. Routing resources are realized by several sub-circuits. Connections between segments are proposed to be realized by the smart via scheme proposed in Sect. 3.4.3.1, and which intends to realize a configuration via between two metal layers. In segment connections, we will consider the simplest case where a single transistor is used between two metal lines. Switch-box cross point nodes will be realized as proposed in Sect. 3.4.3.3. Indeed, a 6-transistor scheme is used for the cross point, while the SRAMs are realized close to the node in a back-end logic scheme. Finally, it is worth noticing that every buffer required (for signals or for clocks), will be realized directly in the back-end, using the logic gates proposed in Sect. 3.4.3.2. This integration is very relevant to our considerations, since it appears inefficient to connect a signal every time to its front-end buffer, before coming back to the metal line.

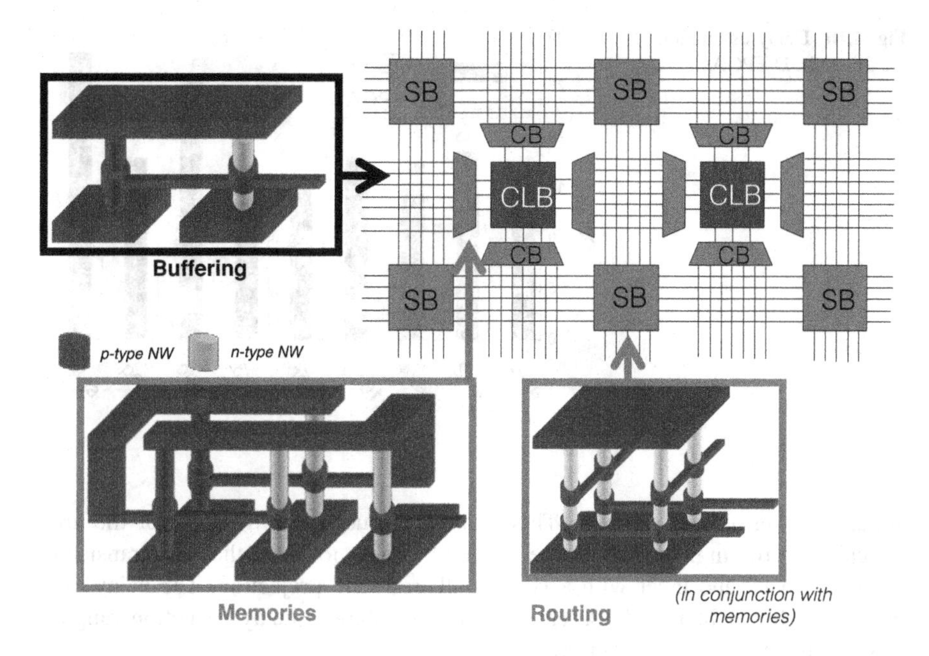

Fig. 4.9 Vertical-NWFET-based FPGA overall organization

In the proposed organization, the connection boxes and logic blocks remain in the front-end levels. Thus, we expect to distribute equally the complexity between the circuits implemented in front-end and those moved in back-end.

Figure 4.9 gives an overview of the final FPGA scheme, based on the Vertical NWFET technology.

4.5.2 Methodology

In this evaluation, the methodology is again the same as that of Sect. 4.3.2 and Sect. 4.4.2. Since the technology "only" improves the routing resources, we adapt the architecture file accordingly. The performance of each elementary block in the back-end uses the numbers extracted from TCAD simulations directly.

4.5.3 Simulation of Large Circuits

Simulations have been carried out by mapping the benchmark to SRAM-Bulk and the vertical integrated FET scheme proposed above. We thus evaluate the area and the critical delay. We measured, relatively to 2D CMOS FPGA, an area saving of

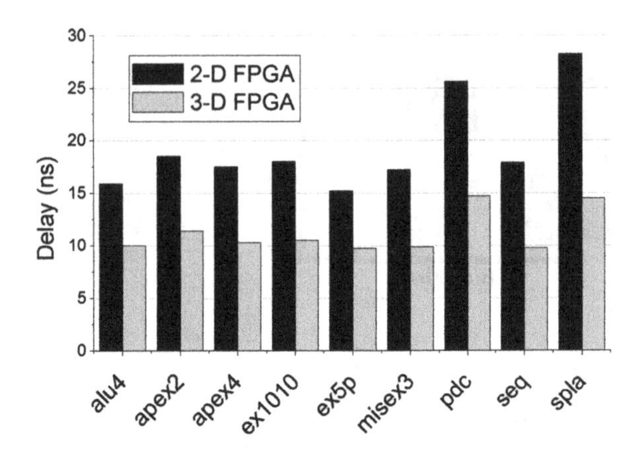

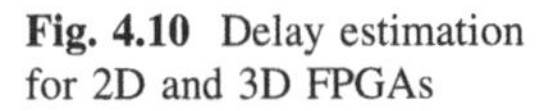
Fig. 4.10 Delay estimation for 2D and 3D FPGAs

about 46% on all our samples. This number is due to the decrease of the area allocated to routing. Indeed, the technology is able to place the pass-transistors inside the back-end layer with a very small front-end projection. The delay estimation is shown in Fig. 4.10. The benchmarks show a delay reduction ranging from 37 to 48%, with 42% on average.

In fact, with the proposed integration process, it is possible to realize large transistors with very low front-end projection and with very good electrostatic channel control. In fact, the Gate-All-Around structure leads to the existence of a channel controllable from all points on the circumference of the wire. As such, for a small diameter d, it is possible to achieve a large effective transistor width (πd).

4.5.4 Discussions

The presented technology is suited for reconfigurable applications, since it yields significant reductions in circuit area and critical path delay. However, only large transistors (>100 nm diameter) are currently manufacturable with this technique. While this first appears to be contrary to the general tendency of scaling, it is in fact not a constraint from a design perspective. In fact, large transistors exhibit good performance levels and can be used as high performance switches, while at the same time they can be realized with a low front-end silicon area. This leads directly to an improvement of area and delay.

The presented results are of interest both for circuit design and for technology. While the technology is not mature enough for mass production, we have used a predictive evaluation methodology to study its impact in a full application environment. The results achieved helps to conclude that it is not always necessary for technology to reach the most advanced scaled dimensions. It is thus preferable to converge through a reliable solution for mass production, in terms of variability,

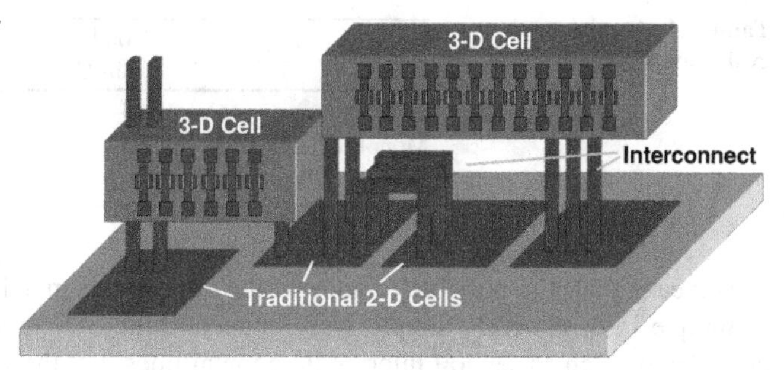

Fig. 4.11 Illustration of the 3D and 2D cell co-integration

clean room compatibility … Feeding back conclusions to the technologists is also a significant achievement for the methodology.

In this thesis, we focus on FPGAs. Nevertheless, it is worth noticing that this proposal should not be limited to this application only; complex 3D ASICs can also be envisaged. Figure 4.11 shows how the technology could be used to stack several layers of standard cells and make them share the same space as routing layers. In this way, routing layers could then not be used entirely and include some areas where routing is not allowed. Back-end logic is a promising technology for numerous applications, but several questions remain open concerning EDA tools with 3D technology. These questions have been heavily researched for 3D TSVs [10]. It is potentially possible to make use of the available tools for coarse-grain circuits and extend them to the fine granularity of this approach.

Finally, we could also expect to use other real 3D technology to fulfill the role of a vertical switch. In this way, it is possible to use vertical *Nano-Electro Mechanical Systems* (NEMS), with nano-relays realized vertically [11]. This technology is even more promising than the FET one, thanks to the very good on/off resistances of NEMS.

4.6 Discussion

Table 4.1 gives a summary of the gain observed through the use of the different technologies. We should note that the best results are obtained by the vertical NWFET technology for the area, with a gain of 46%, and by the PCM resistive memories for the critical path delay with a gain of 44%.

In terms of area, the vertical NWFET technology is the most compact in terms of implementation of routing resources. Indeed, the possibility to move most of the active volume into the third dimension leads to a large reduction in the front-end cell area. The impact on the whole circuit is thus of around 46%. The Monolithic 3D integration process has a similar effect. Nevertheless, it only stacks layers of 2D transistors. Thus, the impact of a cell is much greater than the real 3D implantation. Furthermore, conversely to the monolithic 3D process, the vertical FET integration is able to place the transistors above with a very small front-end impact, but it is also possible to choose the vertical placement of the transistors.

Table 4.1 Architectural evaluation summary

	Total area gain(%)	Critical path delay gain(%)
PCM	13	44
Monolithic 3D	21	22
Vertical NWFET	46	42

This means that it is possible to address the question of metal line layout with this technique in a relatively simple way. In monolithic 3D, some techniques can be considered, such as several intermediate metal lines, but these solutions still suffer from technological issues in terms of process temperature budget. Finally, the PCM gives only an improvement of 13%, which is due to the fact that all PCM circuitry needs to be addressed during its programming. While the PCM programming requires a large amount of current, the remaining access transistor is large, and leads to a loss in terms of gain. Other resistive memory technologies might be envisaged to overcome this limitation.

In terms of critical path delay, the best gain is achieved by the PCM implementation with a gain of 44%, while the vertical NWFET implementation is very close with 42%. Delay in FPGA data paths depends linearly on the on-resistance of the programmable switches, which means that the smaller the resistance, the better the gain in delay. The proposed PCM structure places a resistance memory directly into the data paths. Its on-state resistance is lower than standard silicon transistors, and this leads to the reduction of delay. Vertical NWFET transistors are naturally large, due to their realization in the metal layers. These large transistors exhibit good electrical properties, which makes them able to work as high performance switches, correlated to the on-state of the switch and good delay reduction. Finally, it is worth noticing that monolithic 3D improves the delay by 22%. Even if it this figure is lower than that of the two other technologies, the gain is significant. It is due to the good electrical properties of FDSOI, linked with the reduction of routing wires along the circuit.

Other metrics might be envisaged in the future. In particular, the analysis of power consumption of the whole proposed FPGA should be another comparison objective. However, this requires power models of the architecture and specific tools, such as in [12]. Nonetheless, we could imagine that a solution based on high performance complementary logic, such as vertical NWFET, will lead to the lowest power consumption. On the opposite, RRAM technology is purely passive and will contribute to leakage (impact of low R_{off}) and resistive losses (in the data path).

4.7 Conclusion

In this chapter, we explored the impact on the FPGA architecture of the technologies introduced in the previous chapter. The evaluated technologies are expected to improve the routing resources of FPGA, thanks to the use of the third dimension. In the context of a conventional FPGA architecture, we benched well-known circuit functionalities using the improved structures and compared the results with respect to traditional CMOS counterparts. We have seen that the best results are obtained by the PCM technology for the critical path delay with 44% in gain, thanks to its low on-resistance value. The vertical NWFET technology improves the area by about 46% and the critical path delay by 42% respectively. These improvements are due to the extremely promising performance levels of this high-performance and fully 3D technology. Finally, we should also remark that monolithic 3D integration yields an improvement of 21% in area and 22% in critical path delay. This relatively mature technology thus represents a good trade-off for several performance metrics.

References

1. J. Luu, I. Kuon, P. Jamieson, T. Campbell, A. Ye, M. Fang, J. Rose, VPR 5.0: FPGA CAD and architecture exploration tools with single-driver routing, heterogeneity and process scaling, *ACM Symposium onFPGAs*, FPGA '09, Feb 2009, pp. 133–142
2. V. Betz, J. Rose, A. Marquart, Architecture and CAD for deep-submicron FPGAs. (Kluwer Academic Publishers, New York, 1999)
3. BLIF circuit benchmarks: http://cadlab.cs.edu/~kirill/
4. ABC: Berkeley logic synthesis tool, http://www.eecs.berkeley.edu/~alanmi/abc/
5. Versatile packing, placement and routing tool for FPGA, http://www.eecg.utoronto.ca/vpr/
6. G. Betti Beneventi, L. Perniola, A. Fantini, D. Blachier, A. Toffoli, E. Gourvest, S. Maitrejean, V. Sousa, C. Jahan, J.F. Nodin, A. Persico, S. Loubriat, A. Roule, S. Lhostis, H. Feldis, G. Reimbold, T. Billon, B. De Salvo, L. Larcher, P. Pavan, D. Bensahel, P. Mazoyer, R. Annunziata, F. Boulanger, Carbon-doped GeTe phase-change Memory featuring remarkable RESET current reduction, in *Proceedings of the european solid-state device research conference (ESSDERC)*, pp. 313–316, 14–16 September 2010
7. R. Waser, Electrochemical and thermochemical memories, *IEEE International Electron Devices Meeting*, pp. 1–4, 15–17 Dec 2008
8. M. Kund, G. Beitel, C.-U. Pinnow, T. Rohr, J. Schumann, R. Symanczyk, K.-D. Ufert, G. Muller, Conductive bridging RAM (CBRAM): an emerging non-volatile memory technology scalable to sub 20 nm, *IEEE International Electron Devices Meeting*, pp. 754–757, 5 Dec 2005
9. E. Ahmed, J. Rose, The effect of LUT and cluster size on deep-submicron FPGA performance and density. IEEE Trans. Very Large Scale Integr. Sys. **12**(3), 288–298, March 2004
10. K. Siozios, K. Sotiriadis, V. F. Pavlidis, D. Soudris, Exploring alternative 3D FPGA architectures: design methodology and CAD tool support, *17th International Conference on Field Programmable Logic and Applications (FPL)*, pp. 652–656, 2007
11. J. Rubin, R. Sundararaman, M.K. Kim, S. Tiwari, A single lithography vertical NEMS switch, *IEEE 24th International Conference on Micro Electro Mechanical Systems (MEMS)*, pp. 95–98, 23–27, Jan 2011
12. F. Li, Y. Lin, L. He, D. Chen, J. Cong, Power modeling and characteristics of field programmable gate arrays. IEEE Trans. Comput. Aided Des. Integr. Circuits Sys. **24**(11), 1712–1724, Nov 2005

Part III
Disruptive Logic Design

Chapter 5
Disruptive Logic Blocks

Abstract In this chapter, emerging technologies will be used to create disruptive elements for Field Programmable Gate Arrays. We focus mainly on the combinational function blocks, in order to improve the computing performance of future reconfigurable systems. We propose to study the use of an ambipolar carbon electronics process and two different silicon nanowire crossbar processes. Carbon electronics, and especially the Carbon Nanotube Field Effect Transistor, exhibits the property of ambipolarity, which means that *n*- and *p*-type behaviors are achievable within the same device. It thus becomes possible to obtain a device with tunable polarity, thanks to the addition of a second (polarity) gate to the device. This novel programmability of CNFETs is leveraged in a compact in-field reconfigurable logic gate and in a new approach to designing compact dynamic logic gates. We then propose the use of a sublithographic silicon nanowire crossbar process. It is worth noticing that using the crossbar organization helps to compact the dimensions (up to 6×) required by the logic circuits. Nevertheless, a technological process build around a sublithographic arrangement of nanowires is highly unreliable, and its feasibility remains uncertain when considering all the access contacts. In order to correct the lack of manufacturability of the sublithographic crossbar process, we propose a variant on this crossbar process. This is realized on a modified Fully Depleted Silicon-On-Insulator process, and enables the construction of circuits in a crossbar scheme with lithographic dimensions.

In the previous chapter, we focused on improvements of the routing and memory parts of the FPGA architecture. While these improvements are relevant to solve many issues in FPGAs, the proposals can still be considered only to be incremental improvements of the conventional structure. It is thus of interest to consider the limitations of FPGAs more carefully. Indeed, in an FPGA architecture, only 14% of the area is used for logic blocks, while the entire chip is occupied mainly by "peripheral circuitries" [1].

P.-E. Gaillardon et al., *Disruptive Logic Architectures and Technologies*,
DOI: 10.1007/978-1-4614-3058-2_5,

Logic blocks are the core of computation and thus the architecture could be considered to be inefficient with respect to the ratio between computing and periphery. We will therefore try to explore another direction for the architecture. In this chapter, we will work on the elementary logic block, as it is the foundation of the entire architecture.

5.1 Context and Objectives

As introduced, we focus in this chapter on logic blocks. The main objective is to provide new seeds for basic logic in reconfigurable architectures. These seeds will serve to build new architectural schemes in Chap. 6.

To work in this direction, we intend for each block to be disruptive as compared to conventional technology. We define the disruptiveness as a gain of around 10x with a combination of area/performance/power metrics.

Two principal hypotheses/requirements must be formulated for this goal.

Firstly, the disruptive blocks must be designed in an "architecture-friendly" manner, meaning that the blocks must be adapted to the architectural arrangement. This implies, for example, that a compact block which requires a large amount of extra circuitry is not suited for large applications. To illustrate this aspect, we consider diode logic [2], which needs signal restoration after each computing block. So even if the logic element is small compared to an equivalent MOS, the number of additional output buffers required to ensure correct logic behavior can lead to a significantly reduction in the overall area gain. In the same way, logic blocks should minimize memory requirements since, similarly to the preceding discussion, additional memory leads to an increase in overhead. Further, the memory logic levels should also be suited to the logic.

Secondly, the design must be feasible from a technology point of view, and moreover be adapted to the strengths and weaknesses of the fabrication process. First of all, it seems quite fundamental that the envisaged technologies should remain compatible with standard CMOS processes. It is clear that semiconductor companies must amortize the cost of current existing processes and will not abandon their facilities and equipment. Due to this, any proposal must be compatible with clean room facilities by, for example, avoiding the use of gold as a catalyst. All technological assumptions in this chapter are thus driven by analyses from industry technology department capabilities.

Several emerging devices have been proposed during the last decade, which exploit various physical phenomena and properties. In this chapter, we consider two principal means of improvement of circuit performance: the use of improved device functionality and/or the use of increase in device integration density.

The first direction is based on the improvement of device functionality. Here we consider in particular *Double-Gate Carbon Nanotube Field Effect Transistors* (DG-CNFET), which can be configured to *n*-, *p*- or *off*-type devices depending on the polarity-gate voltage. Such device ambipolarity is of benefit for complex logic designs. Indeed, this property will be explored to build a dynamically reconfigurable logic cell. Then, the use of in-field reconfigurability will be generalized to dynamic logic circuits.

The second direction is based on the improvement of elementary device integration density. Based on realistic assumptions, a regular process can be used to obtain high integration density, through a variety of different processes. We will examine the sublithographic integration of devices and, based on the NASIC approach [3], we will derive it to implement simple cells, rather than complex circuits. The integration of simple cells changes some paradigms of the NASIC approach, but is motivated from a technological point of view. In order to compare fairly the approach with elementary reconfigurable system gates, we use a multiplexer gate as baseline logic cell. The technological assumptions are mainly based on speculative bottom-up arrangements of sublithographic wires. Thus, in order to simplify the hypotheses, we will move in a second approach from crossbars to a sublithographic process integration. An innovative process, derived from *Fully Depleted Silicon-On-Insulator* (FDSOI), will be proposed. A layout methodology for logic cells using the technology will also be described. Finally, an optimization of the technology will be done by considering the global performance of the gate.

5.2 On the Use of Ambipolar Carbon Electronics

5.2.1 Introduction

Since the discovery of *Carbon Nanotubes* (CNT) in the early 1990s [4] and the first isolation of a graphene sheet in 2004 [5], carbon electronics has witnessed a growing interest. In particular, the study of the intrinsic properties of carbon and its potential application to microelectronic devices has been of high interest.

Carbon nanotubes, which usually consist of a single sheet of carbon rolled up to form a seamless tube [6], possess exceptional electrical properties such as high current carrying capability ($>10^9$ A/cm^2) [7] and excellent carrier mobility (9,000 cm^2/V s) [8]. Due to the small diameter ($\sim$1 nm), nanotubes are ideal candidates to provide *One-Dimensional* (1D) electrical transport. Ballistic transport, even at room temperature, has been demonstrated over short distances ($\lambda_{MFP} \approx 700$ nm) [8]. It should however be noted that only carbon nanotubes built from a single layer of carbon (single-wall carbon nanotubes) can behave as semi-conductors. Multiple-wall carbon nanotubes are composed of a number of coaxial single-wall nanotubes. High-performance electronics applications only use single-wall tubes, due to their small size and good semiconducting behavior.

Graphene is a monolayer of carbon atoms in a honey-comb lattice with unique electronic properties [9]. With a high saturation velocity (5.5×10^7 cm s^{-1}) [10], graphene is considered to be a very promising candidate for high frequency applications. In addition, the ultra-thin body thickness of graphene offers ideal 2D electronics channels for the ultimately scaled down device. Recently, cut-off frequencies in the gigahertz range have been demonstrated in top-gated graphene transistors built on exfoliated single-layer graphene sheets [11, 12], and on few-layer graphene grown on SiC substrates [13]. Epitaxial growth has been investigated in particular to produce wafer-scale, high-quality graphene [14]. Such techniques are compatible with all traditional lithographic patterning techniques to build the electronic circuits, following on from traditional circuit fabrication processes. Nevertheless, the graphene material is a semi-metal, meaning that it has no gap and thus no semiconducting behavior. One way to introduce a band gap is to confine electrons by patterning a 2D graphene sheet into a narrow ribbon (<10 nm), better known as a *Graphene NanoRibbon* (GNR) [15, 16]. Another way to modify the band structure of graphene is to stack two mono-layers to form a bi-layer which has a semiconducting band structure with zero band gap [17]. This band gap can be additionally tuned by creating a potential difference between the two layers [18, 19].

In addition to their promising performance levels, it is worth highlighting that carbon technology is ambipolar. This means that a transistor can conduct for both positive and negative gate voltages. Traditionally, this kind of behavior is not suited for micro-electronics applications, where unipolar devices are required. While technologists are working on masking this property, it is also interesting to examine this property at the circuit level to assess opportunities with such ambipolar devices. In [20], a single carbon nanotube field effect transistor with a resistive pull-up to the power supply was used in order to implement a dynamic XOR logic gate. Further, ambipolarity can be controlled by adding a second control gate, such that it becomes possible to obtain a single functionality-improved device with a tunable polarity [21].

In this section, we will focus on carbon technology, and in particular on the control and efficient use of ambipolarity.

5.2.2 *CNT-Based Device Properties and Technological Assumptions*

5.2.2.1 Device Physics

Most *Carbon Nanotube Field Effect Transistors* (CNFETs) studied so far demonstrate an ambipolar characteristic [21], i.e. conduction for both positive and negative gate voltages. Indeed, conversely to a conventional *Metal Oxide Semiconductor Field Effect Transistor* (MOSFET), switching of a CNFET is dominated by the modulation of Schottky barriers formed at the nanotube/metal contacts [22]. Since carbon nanotubes are intrinsically ambivalent, meaning that both electrons and holes can flow through the structure, the height of the Schottky barriers allows the selection of the carrier conduction. At sufficiently negative (resp. positive) gate voltage, the Schottky barrier is sufficiently thinned to enable hole (resp. electron)

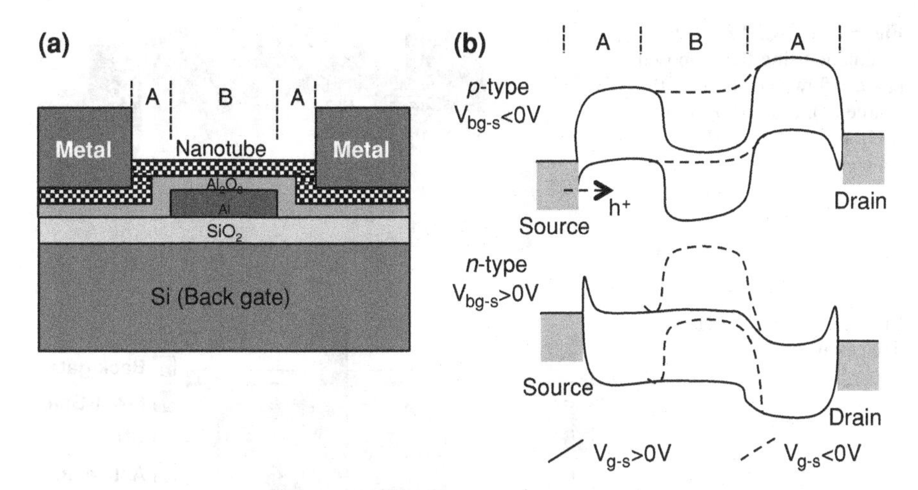

Fig. 5.1 **a** Schematic cross-sectional diagram of a DG-CNFET [21]. **b** Band diagrams of the structure (*Left* $V_{bg\text{-}s} < 0$, and *right* $V_{bg\text{-}s} > 0$) [21]

injection from the source (resp. drain) contact into the nanotube. In the behavior of a FET, these barriers are not desirable because they reduce the controllability of the device. Several works have been proposed to improve the contact quality. In [23], a graphitic wet layer is used to improve the contact between metal and nanotube.

In [21], another approach is taken. Instead of eliminating the ambipolarity of the devices, a dual-gate approach is introduced. The dual-gate creates pure *n*- and/or *p*-type devices with excellent off-state performance, simply by using electrostatic control on the carriers. Figure 5.1a depicts the device cross section of a *Double-Gate CNFET* (DG-CNFET) structure. The DG-CNFET possesses an additional Al gate electrode placed underneath the nanotube between the source and drain contacts. In the design, the Al gate is the primary gate that governs the electrostatics and the switching of the nanotube bulk channel. The Schottky barriers at the nanotube/metal contacts are controlled by the Si back gate (substrate). Using this structure, it has been shown that electrostatic doping effects may be used to eliminate ambipolar characteristics and to obtain a transistor possessing a tunable polarity (*n* or *p*). This *p*-FET and *n*-FET behaviour of the dual-gate CNFET can be understood by the schematic band diagrams shown in Fig. 5.1b.

For a sufficiently negative (resp. positive) Si gate voltage, the Schottky barriers are thin enough to allow for hole (resp. electron) tunnelling from the metal contacts into the nanotube channel. Thus, regions A (Fig. 5.1) become electrostatically doped as *p*-type (resp. *n*-type), resulting in a *p*/*i*/*p* (resp. *n*/*i*/*n*) band profile that allows only hole (resp. electron) transport in the nanotube channel. The dual-gate CNFET is switched *on* and *off* by varying the Al gate voltage that alters the barrier height for carrier transport across region B. In this configuration, regions A serve as extended source and drain, and the device operates similarly to a conventional MOSFET through bulk-switching in region B.

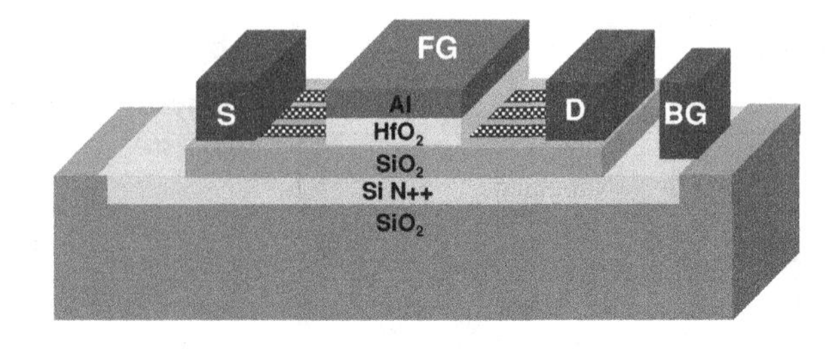

Fig. 5.2 DG-CNFET device schematic using the proposed process-flow and showing the source (S), drain (D), front- (FG) and back-gate (BG) contacts

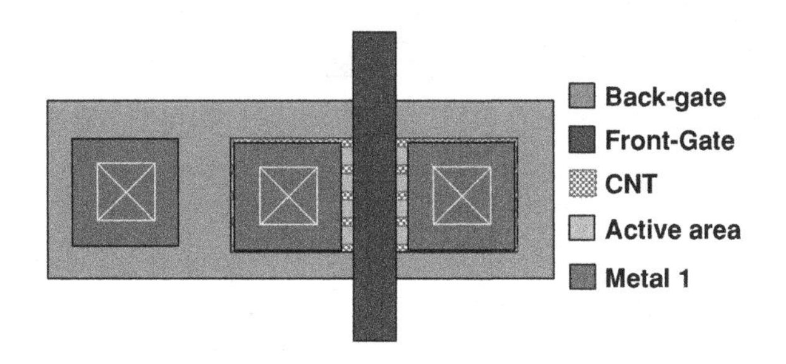

Fig. 5.3 DG-CNTFET device layout

5.2.2.2 Technological Assumptions

In [21], the authors used an integration process based on a generalized and common back-gate electrode. This process is simple for research and characterization purposes, but implies a global and shared back-gate for all the devices. However for circuit design, the principal advantage of the device is its unique in-field reconfigurability, meaning that each back-gate needs an individual control. The proposed process flow ensures individual back-gate access.

The process is based on *Silicon-On-Insulator* (SOI) wafers. In this process, it is possible to build silicon mesas (i.e. islands of silicon surrounded by oxide) to realize the DG-CNFETs back-gate. The individual control requirement will thus be guaranteed. The final device is shown in Fig. 5.2.

The expected technology process is based on a realistic and CMOS-compatible process flow. Starting with a SOI wafer, where the silicon-on-insulator layer will be used as back-gate, *N*++ doping (or NiSi salicidation) can be realized to ensure a good conductivity of the back electrodes. SiO_2 is subsequently deposited as back-gate oxide, and intrinsic carbon nanotubes are transferred on top of this [24]. Then, the gate oxide (HfO_2) and the metal (Al) of the top gate are deposited and patterned. Then, the active area and the back gate are defined by SiO_2 and Si etching respectively. Subsequently, the metal is sputtered onto the contacts to drain, source and both gates.

Using the previously defined process flow, we consider the layout requirements for our DG-CNTFET device. The expected layout is shown in Fig. 5.3. This layout corresponds to parameterized-cell requirements and follows a standard industrial back-end rule set.

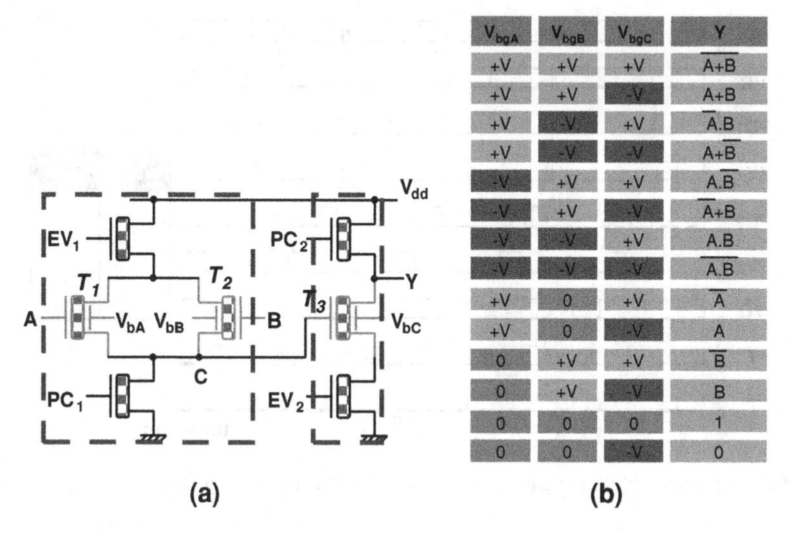

V_{bgA}	V_{bgB}	V_{bgC}	Y
+V	+V	+V	$\overline{A+B}$
+V	+V	-V	$A+B$
+V	-V	+V	$\overline{A}.B$
+V	-V	-V	$A+\overline{B}$
-V	+V	+V	$A.\overline{B}$
-V	+V	-V	$\overline{A}+B$
-V	-V	+V	$A.B$
-V	-V	-V	$\overline{A.B}$
+V	0	+V	$\overline{A}$
+V	0	-V	A
0	+V	+V	$\overline{B}$
0	+V	-V	B
0	0	0	1
0	0	-V	0

Fig. 5.4 **a** Transistor level schematic and **b** configuration table for CNT reconfigurable cell [25]

5.2.3 Functionality Improvement: In-Field Reconfigurability

The in-field reconfigurability property of the DG-CNFET increases the functionality of circuits at the device level. It is thus possible to build compact logic cells using the in-field reconfigurability property. A dynamic reconfigurable logic cell is presented in Fig. 5.4a. The cell is based on dynamic logic and is composed of seven DG-CNFETs organized into two logic stages: logic function and follower/inverter. A four-phase clock signal set is used to perform the logic operation, and consists of two precharge inputs (PC_1, PC_2) and two evaluation inputs (EV_1, EV_2). The signals are non-overlapping as in classical CMOS dynamic logic gates. The polarities (*n*-type/*p*-type) of DG devices T_1, T_2 and T_3 are controlled by the corresponding back-gate bias voltages V_{bA}, V_{bB} and V_{bC}, as previously explained. The cell may thus be configured to one of fourteen basic binary operation modes, as shown in Fig. 5.4b. In this cell, there are seven inputs and one output:

- two Boolean data inputs A and B (The logic levels are represented by the supply voltage values V_{dd} and Gnd);
- three control inputs to configure the circuit according to Fig. 5.4b (Back-gate bias voltages are according to the tuned control values);
- the clocking signals PC_1, PC_2, EV_1 and EV_2;
- the circuit output Y.

Figure 5.5 illustrates how this logic gate works. When $V_{bA} = V_{bB} = V_{bC} = 1$ V, CNFETs T_1, T_2 and T_3 are all configured as *n*-type FETs. When PC_1 is enabled, the first stage is pre-charged, and the voltage of the internal node C (V_c) is

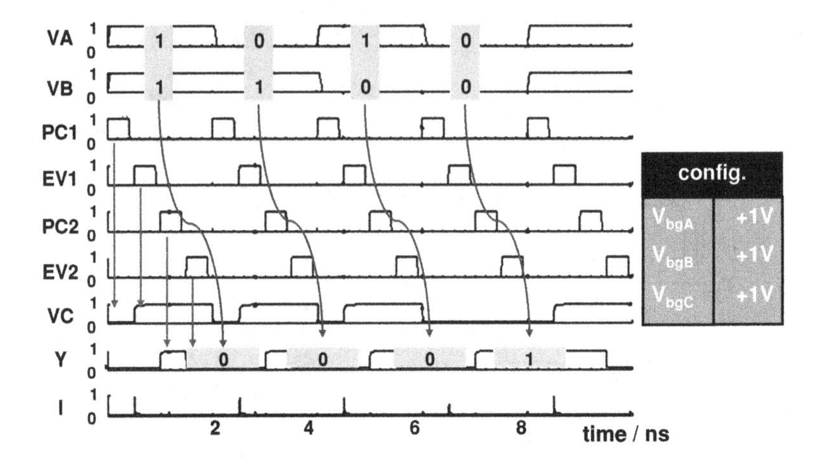

Fig. 5.5 Reconfigurable logic gate waveforms in NOR configuration [25]

discharged to 0 V. If for example either of the data inputs A or B = logic '1', then when EV_1 is enabled, the first layer evaluates its output such that the internal node C is set to logic '1'. Then PC_2 is enabled (pre-charge of the second stage), and the output Y is charged to logic '1'; and when EV_2 is enabled, the output Y is evaluated to logic '0'. In fact in this configuration, the only situation where C is not set to logic '1' and Y therefore evaluates to logic '1' (since T_3 is off) is when both A and B = logic '0'. This means that for $V_{bA} = V_{bB} = V_{bC} = 1$ V, the cell is configured as a NOR gate.

5.2.4 *Performance Evaluation*

The cell presented above promises circuit compactness for reconfigurable applications. We will now assess the performance of the in-field reconfigurable gate and compare it to the equivalent CMOS counterpart.

5.2.4.1 DG-CNFET Evaluation Methodology

In order to evaluate the performance of the DG-CNFET reconfigurable logic cell, we investigate the area, delay and power consumption. The area will be extracted considering its layout in a 22-nm lithographic node. The choice of this extrapolated node is due to the fact that this solution has been estimated for middle-term. Since the logic cell is dynamically clocked, the maximal performance is controlled by the maximal clock frequency reachable. The performance and the power consumption of the DG-CNFET cell were computed by electrical simulations. The simulations are run with the help of a unipolar CNFET model [25].

This approximation is valid since the back-gate voltages remain only DC. However, the model did not take the DG-CNFET intrinsic parasitics into account. Hence, parasitic values were extracted from the layout and the netlist has been corrected accordingly. Simulations have been conducted using different load values for performance evaluation, while a *fan-out-of-four* (FO4) load has been chosen for the power consumption evaluation.

5.2.4.2 Silicon CMOS Performance Evaluation Methodology

In order to evaluate in an objective manner the performance of the advanced DG-CNFET technology, we will consider the use of an advanced lithography node of 22 nm (i.e. the expected equivalent silicon technology node that will be available when CNT technology will be mature enough for industry). To perform this evaluation, we will extract the metric values from a well-established 65 nm technology and extrapolate it, as explained in Table 5.1.

Considering reconfigurable logic architectures, the cell is compared with a standard LUT4:1 as used in actual FPGA systems. A standard MUX4:1 is considered to implement the equivalent CMOS circuit. The performance metrics of the MUX4:1 will be used to illustrate the methodology.

5.2.4.3 Simulation Results

Table 5.2 summarizes this comparative study. The estimated size of a CMOS MUX at the 22 nm node is 1.19 μm^2, while the DG-CNFET cell area is 0.39 μm^2. This gain of 3.1$\times$ can be explained by the significant reduction in the number of transistors, from 26 for a standard MUX to only seven with the in-field reconfigurable cell.

With regards to timing, the DG-CNFET cell demonstrates lower performance values than the CMOS equivalent cell (with all parasitics considered). This is counter-intuitive, considering the intrinsic properties of carbon electronics: (i) that the good electrical transport of carbon should increase the device speed, and (ii) that the size reduction of the elementary cell should reduce the parasitics. While both factors should lead to an improvement in the performance metrics, the comparison in fact considers a static (CMOS) logic cell and a dynamic (DG-CNFET) logic cell. The comparison is thus not completely valid; however it is justified by the fact that dynamic logic is used to compact the cell and to compete with standard FPGAs. Nevertheless, this does lead to an increase time delay, due to the four clock phases used to charge and discharge the capacitive nodes.

This could be improved by using transistors with much higher drive strength. In fact, in the simulation, the DG-CNFET transistor is built with a single nanotube as the channel material. In this context, it seems reasonable to envisage the use of multiple CNTs per transistor, where each CNT contributes to the transistor

Table 5.1 CMOS metrics scaling methodology

Area	Area is measured using the layout view, and then extrapolated to the target 22 nm technology node. The scaling factor used is $K_{Area}(x \text{ nm} \rightarrow \text{y nm}) = \frac{\text{x}}{\text{y}}$ In the 65 nm → 22 nm case (i.e. 3 technological generations following a scaling factor between generations of $\sqrt{2}$): $K_{Area} \approx 2.95 \approx \left(\sqrt{2}\right)^3$ and area is scaled down according to K_{Area}^{-2}.
Performance	To evaluate the maximal achievable performance, the intrinsic delay and load influence factor of the operator must be extracted with respect to the parasitic devices. Estimation of the delay in CMOS MUX is done using a 65 nm standard cell, and the data are then extrapolated to the 22 nm node. The scaling factor used, based on data taken from the ITRS [48], is: $K_{Delay}(x \text{ nm} \rightarrow y \text{ nm}) = \frac{ITRS_{On-Chip\ Local\ Clock}(x \text{ nm})}{ITRS_{On-Chip\ Local\ Clock}(y \text{ nm})}$ In the 65 nm → 22 nm case, the factor used is $K_{Delay} \approx 1.95$, and delay is scaled with K_{Delay}^{-1}.
Power consumption	The predictive evaluation of power consumption evaluation depends on the target load, the associated architecture, configuration details and frequency. An FO4 load has been chosen for the purposes of evaluation. The procedure for the evaluation of CMOS needs extrapolation. First, an evaluation is performed, through simulation, using a 65 nm standard cell at a frequency compatible with this technology and normalized with respect to the frequency. $P_{Hz}(x \text{ nm}) = \frac{P_{Eval}(x \text{ nm})}{f(x \text{ nm})}$ Then, this normalized power figure is scaled to the extrapolated node. $P_{Hz}(y \text{ nm}) = \frac{P_{Hz}(x \text{ nm})}{K_{Power}(x \text{ nm} \rightarrow y \text{ nm})}$ The determination of the scaling factor is quite complex, due to the lack of data on power evolution. In fact, power consumption of a circuit can be split into three contributions: the dynamic power (α.f.C.V_{dd}^2), the static power ($V_{dd} \cdot I_{leak}$) and the short-circuit power. These contributions follow different trends which do not evolute according to the Moore's Law anymore. Nevertheless, while it is clearly impossible to consider that this factor is close to the area scaling factor for such an advanced technology node, some degree of correlation can be found with the frequency scaling factor, evaluated using the ITRS. We thus (somewhat arbitrarily) assume that $K_{Power} \approx K_{Delay} \approx 1.95$ Finally, denormalization to the targeted frequency is computed $P(y \text{ nm}) = P_{Hz}(x \text{ nm}).f(y \text{ nm})$

channel. Using multi-channel transistors will result in a higher I_{on} value, which accelerates the charge and discharge of all the gated stages (with of course corresponding impact on power consumption figures).

Table 5.2 Global evaluation of the DG-CNFET cell performances

	Area (μm^2)	Intrinsic delay (ps)	K_{Load} (ps fF^{-1})	Average power at 4 GHz (μW)
MUX MOS	1.191	79	5.7	3.56
DG-CNFET	0.39	149	124.5	1.78
Gain	×3.1	×0.53	×0.04	×2

Regarding the power estimation figures, we can observe that the cell improves the figures by an average of 2× compared to CMOS.

Regarding the two opposing results (performance and power), we must consider the *Power-Delay Product* (PDP). The PDP is improved by only 6% compared to CMOS technology. In fact, the power consumption is essentially due to the dynamic power contribution. Hence, it depends mainly on the frequency and the load (and parasitic capacitances). Due to the lithographic size of the transistor, the various capacitances of the devices are mostly the same, and this leads to an equivalent PDP.

5.2.5 Dynamic Logic Circuit Improvement

The DG-CNFET ambivalence property has been exploited in previous work to add some reconfigurability to logic cells, while transistor count is maintained low. In this section, we propose to extend the use of the back-gate reconfigurability and its associated states (*n*, *p*, *off*) to improve the structure of dynamic logic cells.

5.2.5.1 Concept

Compact logic cells can be realized using dynamic-logic [26]. This allows, as in conventional CMOS technology, to reduce the number of transistors (and therefore the complexity) by a factor of almost two for each function, with respect to static-logic implementation. This reduction can be explained by the elimination of the *p*-type pull-up network and replacement around the pull-down network by two clocked transistors for precharge and evaluation. The generic scheme for dynamic logic is shown in Fig. 5.6a.

The DG-CNFET *off*-state is used to reduce complexity. The *off*-state corresponds to a transistor that is never *on* whatever the state of the front-gate is. We will use this property to merge the evaluation transistor directly with the function paths.

The programmable polarity allows us to eliminate all input inversion circuitry. Effectively, the polarity of the transistor might be chosen in order to select whether the front-gate signal should be considered directly or as its complementary value. The generic structure is shown in Fig. 5.6b. The information of interest is the

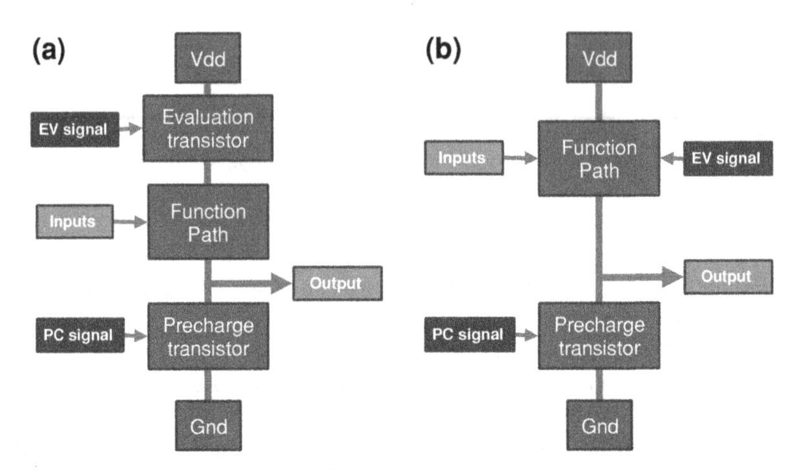

Fig. 5.6 **a** Generalized scheme for CMOS dynamic logic. **b** Generalized scheme for double-gate-based cells

Pull-up (resp. pull-down for the case of the inverted function) network with the evaluation clock merged on the back gates. The function path structure remains the same as in traditional CMOS, and consists of several DG-CNFETs. The DG-CNFETs could be placed in serial branches (for AND functions), parallel branches (for OR functions) or any combination of serial and parallel branches. The evaluation clock signal controls the back-gate of at least one DG-CNFET for each path from V_{dd} to the Output node. The considered DG-CNFETs receive non-complemented inputs on their front gates. In the case of complemented inputs, DG-CNFETs have to be configured to *p*-type using the associated voltage on their back-gate.

5.2.5.2 Buffer Cell Illustration

Figure 5.7a shows an illustration of the concept with a buffer implementation. The cell operation is based on two DG-CNFETs. The precharge transistor (PC-gate, ground path) is configured to *n*-type, via the V_{dd} voltage applied to its back-gate. The evaluation and signal transistor (IN-gate, V_{dd} path) are combined; the evaluation clock signal is connected to the back-gate and the input signal is connected to the front-gate.

Figure 5.7b shows the associated waveform for the cell. To explain the behavior of the cell, three phases are clearly identifiable:

- **Precharge**: PC = 1, EV = 0. The *n*-type bottom transistor is *on*, the top transistor is *off*. The output node is thus precharged to 0.
- **Sleep**: PC = 0, EV = 0. Both the *n*-type bottom transistor and the top transistor are *off*. Any existing charge is maintained on the output node.

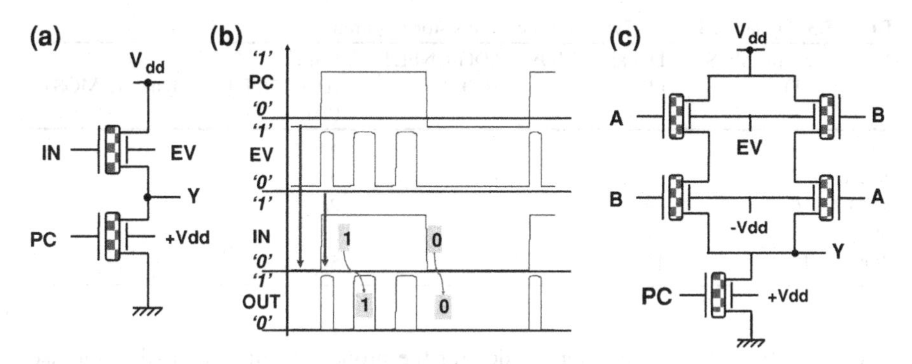

Fig. 5.7 **a** Buffer cell schematic, **b** associated waveform and **c** XOR cell

- **Evaluation**: PC = 0, EV = 1. The n-type bottom transistor is *off*, and the top transistor is in the n-state, allowing evaluation. In this configuration, only the input signal has an influence on the output. If in = 0, then the transistor is off and the output node is unchanged (from precharge state 0). If in = 1, then the transistor is on and output node is charged to 1.

The resulting behavior is a buffer operator working in dynamic logic. It is worth noticing that this architecture may be used to implement all the standard logic operations: Fig. 5.7c presents a XOR gate implementation as another example.

5.2.5.3 Density Improvement Example

Table 5.3 shows a comparison between the proposed approach and its CMOS counterparts in dynamic logic implementation. The proposed implementation, using the DG-CNFET device, reduces the number of transistors up to 50% compared to the dynamic MOS implementation in a standard cell ASIC approach.

5.2.6 Discussion

All the results found above are highly promising from a design perspective. We observe that the functionality improvement of the device reduces the number of used transistors and leads to a more compact logic. Furthermore, the good expected performance levels of carbon electronics improve the power numbers, and certainly the delay also, after some structural improvements. This compactness opens the way towards ultra-fine grain computation, i.e. where the elementary logic blocks will be much smaller than conventional blocks and peripheral circuitries. This will be useful for new architectural paradigms.

Nevertheless, we should also consider that the logic compactness has a high cost in terms of clock and configuration signals: at least four non-overlapping

Table 5.3 Dynamic DGCNFET-based cell transistor requirements

Area	Static MOS (T)	Dynamic MOS (T)	DG-CNFET (T)	Gain (DG-CNFET vs. dynamic MOS) (%)
Buffer	4	3	2	33
Inverter	2	3	2	33
AND	6	4	3	25
OR	6	4	3	25
XOR	12	10	5	50

clocks and three ternary memory nodes for the proposed cell. This makes the use of the cell quite difficult from an architectural point of view. Improvements of the cell using static logic implementations are highly considered for any valid architectures.

5.3 On the Use of 1D Silicon Crossbars

In the preceding chapter, we have investigated a technology that improves the intrinsic functionality of the device and saw how it can lead to a very compact logic cell. In this part, we will focus on the use of a density-improved technology. The dense integration of devices is expected to lead intrinsically to the realization of very compact logic circuits.

5.3.1 Introduction

For several decades, the micro-electronics industry has been known to scale the transistor dimensions, in order to improve the performance. Nevertheless, this increase of performance is not only due to the scaling, but also due to several structural improvements.

Figure 5.8 shows the evolution of the silicon structures from bulk to nanowires. 1D structures demonstrate good electrostatic controllability, which is fundamental to ensure good performance levels for scaled devices. While the trend in industry is towards the *Fully Depleted Silicon-On-Insulator* (FDSOI) process, we can note that nanowires tend to the optimal solution, merging a good electrostatic control with a low *Off* current ($I_{off} < 1$ nA/μm) [27].

Furthermore, in addition to their good expected properties, we can note that the 1D structures make the active dimensions very small. This leads to potentially high-density integrated devices, especially when a crossbar organization is proposed. Thanks to bottom-up fabrication processes, nanowires are useful to build regular crossbar structures at sublithographic scales and open the way towards

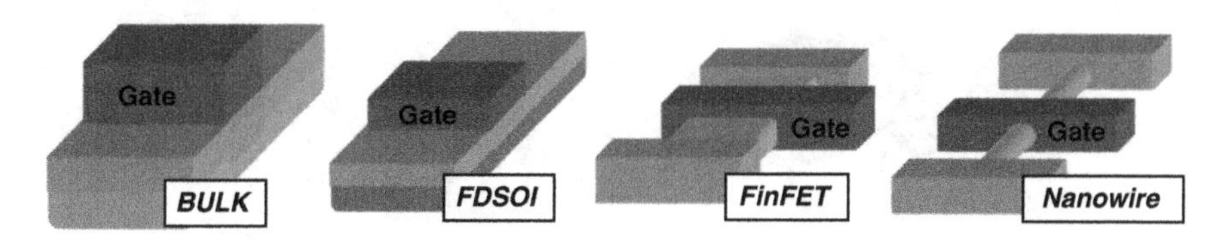

Fig. 5.8 Silicon electronics evolution from bulk to nanowires

enormous device density improvements over CMOS. In the literature, different active elements have been envisaged at the cross points, such as *p–n* junctions [28], molecular programmable switches/diodes [2] or FETs [29]. Several architectures have been proposed, based on these different devices. In [2] molecular diode-switches at crossbar intersections are proposed. This structure forms a diode logic grid, performing Programmable Logic Array. However, diode logic requires level-restoring circuitry and addressing of individual points, which leads to complex interfaces. Considering these limitations, solutions based on FET logic, as introduced in [30], are explored. The NASIC extends this concept in [31] by creating double-stage combinational logic on *Crossbars of NanoWires FETs* (CB-NWFETs). This structure is used as a general fabric to implement logic elements dedicated to nanoprocessors. We will use this approach to implement a logic operator with very compact dimensions.

5.3.2 Technological Assumptions

The crossbar can be manufactured with silicon nanowires grown by Chemical Vapor Deposition, using metallic nanoparticles as catalyst and the Vapor–Liquid–Solid mechanism [32]. This technique allows the achievement of nanowires with diameters controlled by the catalyst size [33] and diameter values around 3 nm [34]. In situ doping during nanowire growth can be used to obtain *n-* or *p*-type nanowires [35]. Nanowires can subsequently be thermally oxidized to obtain a core-shell structure [36] and deposited on a substrate (Fig. 5.9a). The well-known Langmuir-Blodgett technique [37] can be used to align the nanowires and to obtain a first layer of nanowires with spacing controlled by the oxide thickness of the shell [36]. For better alignment when nanowires have very small diameters (i.e. below 3 nm), their lengths cannot be longer than a few micrometers to avoid gradual bending observed on nanowires with these diameters [34]. Nanowire pitch has also to be equal to (or greater than) the corresponding photolithography half pitch of a given technology node plus the nanowire radius for subsequent photolithography steps. The oxide shell should be removed except along the bottom contact between the nanowires and the substrate, in order to avoid removing the nanowires at the same time (Fig. 5.9b). Salicidation of some parts of the nanowires can be achieved using nickel (or platinum) physical vapour deposition,

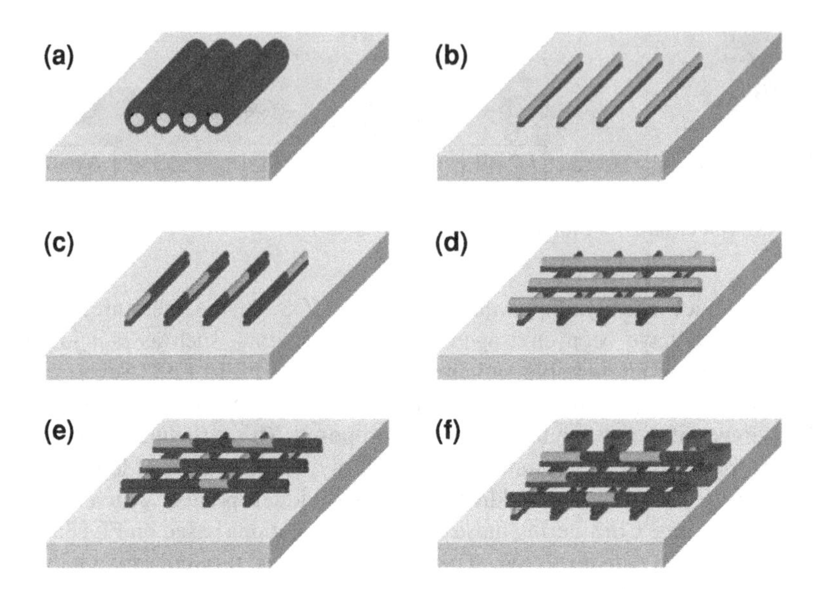

Fig. 5.9 Manufacturing process to build a nanowire crossbar. **a** Deposition of silicon nanowire after growth, oxidation and Langmuir-Blodgett alignment. **b** Etching of the oxide shell. **c** Salicidation of the nanowire regions which will not be transistor channels. **d** Deposition, alignment and etching of the oxide shell of the second layer. **e** Salicidation as step c. **f** Metallization to contact the nanowires around the crossbar

photolithography and etching to define the regions where the silicon nanowire will be the channel of transistors, annealing to form NiSi (or PtSi) regions and chemical removal of the unreacted deposited metal [29] (Fig. 5.9c). The second array of nanowires (crossing the first) can be obtained with the same process sequence (Fig. 5.9d). The oxide shell along the bottom of the second layer of nanowires serves as the gate dielectric and the NiSi (or PtSi) region serves as the metallic gate in the MOSFET structure (Fig. 5.9e). In the same way, some nanowire regions of the second layer can serve as the channel of the MOSFET, controlled by the nanowire metallic regions of the first layer. The process ends with a final metallization to contact all nanowires around the crossbar area by a conventional sequence of deposition, photolithography and etching steps (Fig. 5.9f).

5.3.3 SiNWFET Configurable Logic Cell

5.3.3.1 Nanowire Crossbar Logic

The previously presented process flow is able to build arrays of nanowires. As suggested in [30], FETs are built at the cross points. It is then possible to build elementary combinational logic functions, such as NAND, NOT or Buffer, with a dynamic-logic design style. An example of a dynamic NOT function is given in Fig. 5.10. Precharge and evaluation transistors are used, driven by two non-overlapping clock signals. The associated waveform has been obtained by simulation using an elementary model, taken from [30]. The logic stage is first precharged to V_{dd} followed by a conditional evaluation with respect to the inputs. Thus, it is possible to build every logic function. Nanowire crossbar logic has been widely envisaged at the architectural level. In [3], the concept is generalized with the *Nanoscale Application Specific Integrated Circuit* (NASIC). The NASIC is an architecture intended to realize complex and specific designs within a lithographic crossbar framework.

5.3.3.2 Multiplexer Design Methodology

To ensure an equivalent behavior between this approach and the previous one, we propose the use of a 4:1 *Multiplexer* (MUX). The considered configurable logic cell is designed to fulfill sixteen basic binary operations, at any given time. The selection of operation (i.e. the configuration of the cell) is made through the use of binary configuration signals. In fact, this corresponds to work with the functionality of a *Look-Up Table* (LUT), with no considerations on memories, since we focus on the logic part.

The result is a logic equation which can be implemented in canonical form (with I representing inputs):

$$\sum\prod(I,\bar{I}).$$

The MUX logic function is expressed and simplified through the Espresso logic minimizer tool [38].

A crossbar using only *n*-type NWFETs is able to implement NAND–NAND logic easily [31]. Considering the following property

$$\sum\prod(I,\bar{I}) \equiv \overline{\prod\overline{\prod(I,\bar{I})}},$$

any logic operator could be implemented in two NAND stages. The NASIC approach implements NAND/NAND logic for realizing complex circuit implementations, such as microprocessors (WISP-0) [39]. Due to the complexity, several NASIC tiles are connected. The connection is realized with nanowires.

In our implementation of MUX, we cannot use nanowires for interconnections between MUX and other cells. This assumption comes from the technology. In fact, it is not possible to realize wire alignment when the aspect ratio is too high (i.e. their lengths cannot be longer than a few micrometers to avoid gradual bending observed on nanowires with small diameters [34]). Thus, we expect that long connections will be implemented by micro-scale wires. In summary, the NASIC approach is used to build elementary gates in our approach. To minimize inherently area-hungry connection requirements at the micro-scale, the external contacts for complementary inputs are rendered superfluous through the use of an

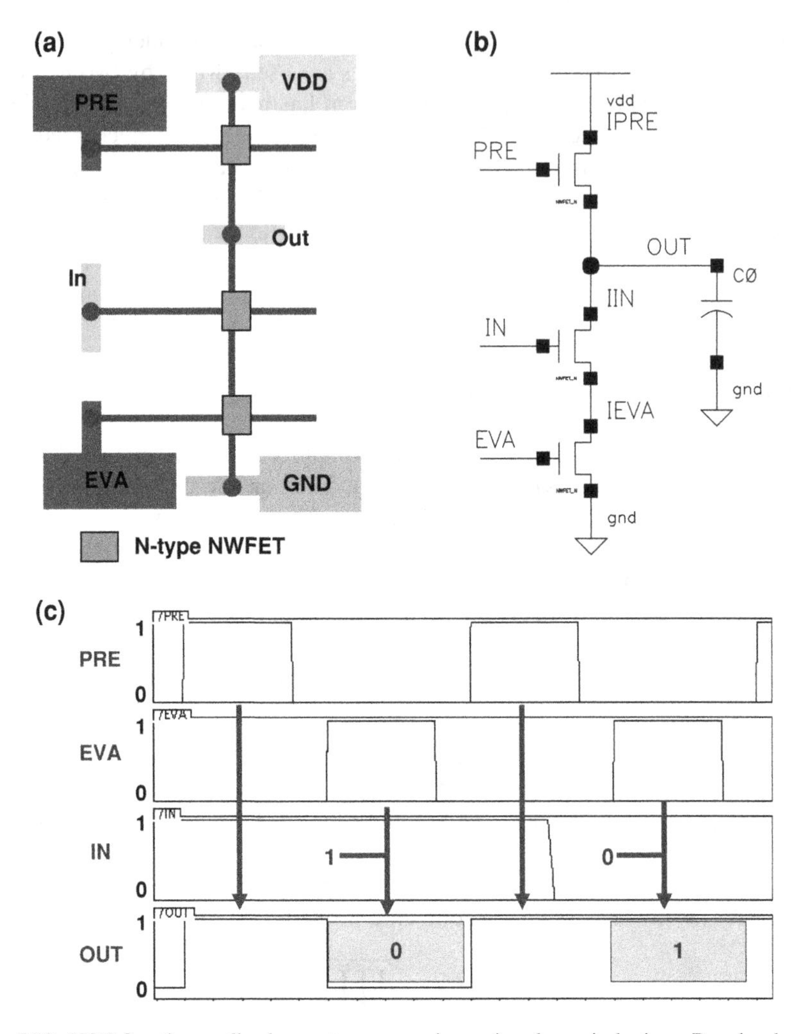

Fig. 5.10 NOT function realized on *n*-type nanowires using dynamic logic. **a** Pseudo-physical view, **b** Schematic view, **c** associated waveform

embedded inverter stage which is placed before the logic. A buffer stage is also coupled in order to ensure the data synchronization required in dynamic logic.

Figure 5.11 shows a representation of the logic cell. The structure is formed by cascading three stages:

Input → NOT/Buffer → NAND → NAND → Output

The cell uses only two sets of dynamic clocks in order to reduce the number and complexity of clock wires. The three pipelined stages operate sequentially using these two clock phases: the NOT/Buffer and output NAND stages operate in phase, whereas the internal NAND stage operates on the opposite phase, ensuring correct charge transfer through the structure.

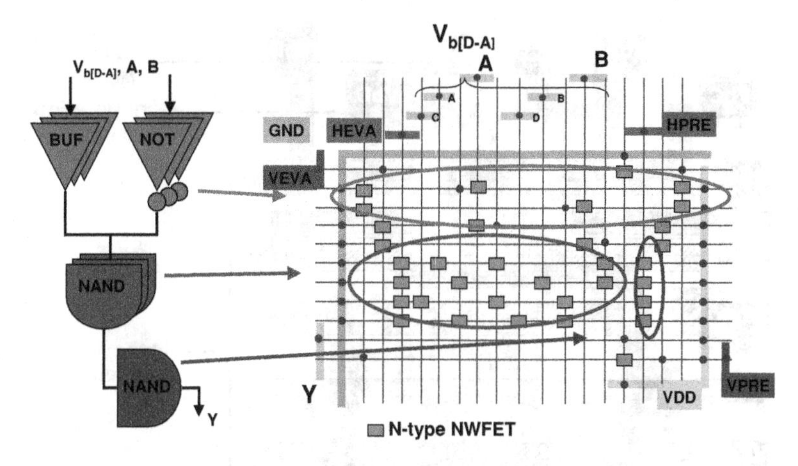

Fig. 5.11 Representation of CB-NWFET dynamic reconfigurable logic cell

Due to the crossbar structure and the fabrication processes, the NAND gates are not all identical. For the first NAND stage, the inputs are y-axis nanowires, and outputs are x-axis nanowires, while the second NAND stage is exactly the opposite. Thus, due to the structure, transistors must be formed on two levels.

5.3.3.3 Layout Design

Small crossbars are connected directly to the microscale, in order to interconnect cells using typical back-end technologies. Nanowires are contacted to typical metal interconnections. It is thus possible to proceed with back-end design and subsequent Design Rule Check. An example layout, which respects an industrial back-end Design Rule set, is shown in Fig. 5.12.

All service signals (clocks and power) are placed on horizontal lines to be shared with other cells. The configuration signals (V_{bA}, V_{bB} …) can be found on large vias to upper metal layers. This is of particular interest for cell addressing or back-end memory implementation. Finally, the data signals (A, B and Y) are available on lower level pads to realize local interconnections with neighboring cells.

5.3.4 Performance Evaluations

5.3.4.1 Methodology

In order to evaluate the performance of this approach, we investigate the area, delay and power consumption. To ensure a fair comparison with the previously

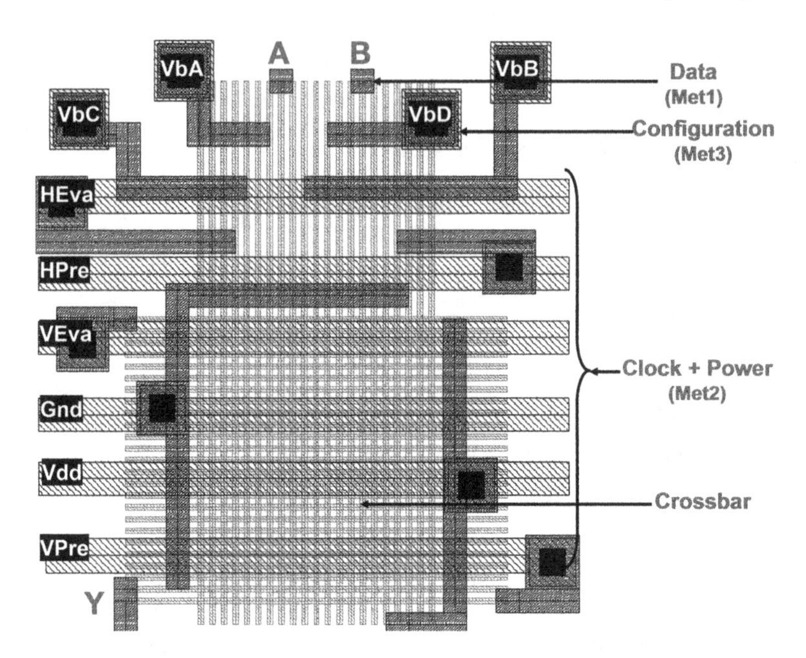

Fig. 5.12 Metal/via layout of the dynamic logic cell

presented approach, we will use the same methodology and hypotheses. The area will be extracted from the proposed layout in the case of a 22-nm lithographic node. Since the NWFET cell is based on dynamic logic, the delay measurement will determine the minimum possible on-time pulse for the clock signals. Parasitic capacitance values have been extracted using the layout implementation and simulations have been conducted under various load value conditions. The elementary model of NWFETs presented in [30] has been used with electrical simulation.

5.3.4.2 Simulation Results

Table 5.4 summarizes this comparative study. Due to the sublithographic dimensions, the solution based on CB-NWFET gives an improvement in density of 4.1×, compared to the extrapolated CMOS.

Regarding the delay analysis of the sublithographic multiplexer circuit, we may observe that the intrinsic delay is close to 17 ps, while the K_{Load} factor is close to 15.1 ps fF^{-1}. By comparison with the data extracted from CMOS technology, we can observe an improvement of 4.6× on the intrinsic delay, while the load factor is degraded by 0.38×. The intrinsic delay depends on the internal data paths. Due to the sublithographic dimensions, it has been possible to reduce the internal parasitic element values, thereby improving the delay figures. Nevertheless, while the

Table 5.4 Global evaluation of the CB-NWFET MUX performances

	Area (μm^2)	Intrinsic delay (ps)	K_{Load} (ps fF^{-1})	Average power at 4 GHz (μW)
MUX MOS	1.191	79	5.7	3
CB-NWFET	0.289	17	15.1	6.7
Gain	×4.1	×4.6	×0.38	×0.45

internal delay is improved by the small dimensions, it is worth noticing that the circuits driving the output are also small. This means that the output buffer is quite inefficient for driving the load capacitance, and thus explains why the load factor is decreased conversely to CMOS equivalent. The output drivers could be improved by parallelizing the paths to increase the *on*-current.

Power consumption figures are clearly worse than those of CMOS by, on average, a factor of 2.2×. It is worth noticing that the comparison is made between a dynamic and a static logic cell, which is not a favorable case for the former. For example, in the case of function '0', only leakage current contributes to power consumption for the static cell, whereas the dynamic implementation charges and discharges nodes continuously. Another point to bear in mind is the model used. This model, with a simple variable resistance, is not favorable for NWFET results, due to a large resulting leakage current. Hence, it will be of high interest to re-run the power figure evaluation using a more realistic model, such as the one proposed in [40, 41].

5.3.4.3 Discussion

As shown by the results, the solution is of high interest for area improvement and intrinsic delay reduction. The sublithographic organization helps to increase drastically the integration density of logic elements, while the sublithographic dimensions also reduce the parasitic contributions. The internal stages of the structure are fast, due to this reduction in parasitics, and the intrinsic delay is consequently improved.

Nevertheless, the contribution of sublithographic dimensions does have disadvantages. It is worth noticing that, even if the computation part is made more compact, several connections to external signals are required. This assumption is motivated by the technology and the difficulty to implement long nanowire interconnections. Due to this, metal lines follow lithographic dimensions, such that connections from the microscale to the nanoscale occupy a large part of the circuit.

Furthermore, as regards performance levels, the output drivers are built with sublithographic devices, which mean that they have a low drive capability. This can be seen from the load factor, which is high conversely to CMOS.

Finally, the main obstacle of this solution remains the technological process flow. Circuit fabrication is based on a bottom-up organization of nanowires. This process is currently in the early stages of its development, which makes the proposed solution highly long-term and speculative.

5.4 On the Use of Lithographic Crossbars

The previous proposal exploits the large density of active elements, built with a sublithographic integration process. Density-based technologies are obviously well-suited to large area reduction. Nevertheless, they are often based on highly speculative technology assumptions. In this section, we will propose a novel integration process for dense crossbar arrangements, based on lithographic FDSOI. Layout and specific methodology will be described, along with circuit performance-driven optimization of the technology.

5.4.1 Introduction

Previous research works addressing the crossbar architecture generally considered bottom-up nanowire fabrication techniques. The low maturity of the technology implies several limitations on the assessment of the architecture. On the one hand, bottom-up fabrication techniques yield dispersed nanowires with highly variable position and spacing. Thus, mean values with respect to the geometry have to be assumed in order to characterize the architecture. On the other hand, the ability to define two layers of perpendicular nanowires is a very challenging technological task. To date, it has been demonstrated only for metallic nanowires in a top-down approach [42], while silicon-based nanowire crossbars are dispersion-based, micrometer-scale demonstrators without any functionalization of the cross points [43]. Capacitive coupling between parallel nanowires has been addressed only in routing applications [44].

In this proposal, we will assess the questions related to the technology limitations and to the impact of the coupling effects between semi-conducting nanowires. Our approach is based on the choice of a fully characterized industrial technology. We propose a process flow to fabricate lithography-based crossbars with FETs as active devices at the cross points. Then, we electrically simulate the circuit in order to assess its performance and the potential loss due to the capacitive coupling and the resistance of the long nanowires.

5.4.2 Technological Assumptions

Using a *Fully-Depleted Silicon-on-Insulator* (FDSOI) process, wires can be manufactured with ultra-regular lines as demonstrated in [45]. In this section, we conceptually complete the already established process for a single layer of parallel nanowires, with a perpendicular top layer of parallel nanowires. In the proposed process, the bottom-most nanowires are defined using photo-lithography at the lithographic pitch, where their dimensions can be controlled through oxidation and etching below the lithographic limit down to 15 nm [46]. Thereby, both *n*- and *p*-type dopings are allowed. On the other hand, the topmost lines are defined as *polycrystalline silicon* (poly-Si) stripes at the lithographic scale. These two perpendicular layers of parallel lines form a crossbar whereby the intersections are called cross points. In such a crossbar, the top lines can electrostatically control the nanowires underneath at the cross points in a FET fashion, when the ladders are covered by a gate oxide. Moreover, the top nanowires can form an ohmic contact to those lying underneath when a via is defined at the cross point.

Figure 5.13 shows the associated process flow. A *p*-type SOI substrate is patterned by lithography to form parallel ridges that are subsequently etched into nanowires. *Plasma Doping* (PLAD) is used to softly define *n*-type wires [47]. Then, the nanowires are passivated in oxide and planarized (Fig. 5.13a). Following this step, the passive regions, i.e., the parts of the nanowires connecting every series FET, are defined by *n*- and *p*-type PLAD on the *p*- and *n*-type nanowires respectively (Fig. 5.13b). It is worth noticing that the implantation step is performed softly because of the small dimensions of the nanowires, such that dopant migration is limited. Moreover, the nanowires are separated by an oxide, further limiting the diffusion of dopants. This reduces the requirements on spacing, which are generally included in the design rules. This allows the smallest lithographic dimensions for all operations of patterning and doping to be reached. Then, the gate stack is defined by depositing the gate insulator, followed by the poly-silicon gate deposition and etching steps (Fig. 5.13c). The poly-silicon lines carrying the gates are defined with regular parallel lines. At this level, the active devices are defined and the east–west connections between them are established through the passive parts of the nanowires, operating as resistances.

The north–south connections are composed of the poly-silicon lines and require the definition of vias between them, as well as the nanowires underneath them. The vias are defined by etching the poly-silicon lines and filling them with metal. In order to decrease the resistance of the north–south poly-silicon lines and the passive parts of the east–west silicon NW lines, it is possible to sputter a thin layer of nickel (or platinum) over the whole structure, which diffuses into the silicon and poly-silicon and forms a low resistance silicon silicide (Fig. 5.13d). For this reason, it is important to first etch the oxide covering the passive regions before sputtering the metal. The remaining metal after the diffusion can be removed by wet etching [30]. Finally, the contacts between the crossbar and the outer circuit are implemented through conventional metallization steps (Fig. 5.13e).

5.4.3 Logic Design Methodology

The FDSOI crossbar technology presented in the previous section can be used to implement different logic gates with a very high active area density.

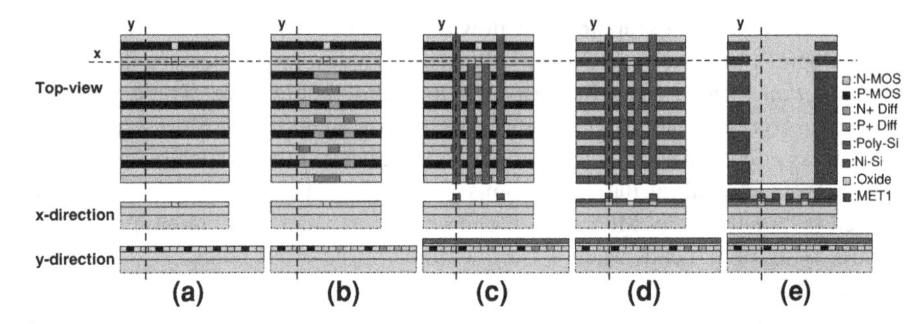

Fig. 5.13 Manufacturing process to build a FDSOI crossbar. **a** Grating patterning and active regions doping. **b** Passive regions definition. **c** Gate deposit. **d** Passive regions finalization and salicidation. **e** Metallization to contact passive regions

5.4.3.1 Inverter construction

An inverter can be fabricated for instance as depicted in Fig. 5.14 with very dense active areas. In this example, p-type and n-type transistors are realized with two separate wires, but aligned to efficiently share the gate on the same poly-Si line. It is worth noticing that the width of p-type transistors could be different from the width of n-type transistors, since separate lines are used for n- and p-type channels. However, in a context of crossbars with ultra-regular layouts, it is preferable to keep all dimensions equal and at the lithographic limit ($L_n = L_p = W_n = W_p = F$), despite the clear impact on skewed delay.

5.4.3.2 Generalization to Complex Logic Circuits

Using the above strategy, it is possible to build any conventional logic function by folding complementary branches around the same contact pad. In the proposed technology, doping of adjacent lines is controlled only by lithography. It is possible to alternate the n- and p-type dopings at adjacent wires or group of lines. This is useful for design purposes, when it is required to make islands of p- and n-transistors or to distribute them regularly.

- Separate Doping Regions

This patterning style consists of grouping p-type and n-type regions separately. This is close to the traditional representation of CMOS circuits and suitable for circuits where the output is a common pin for all branches. This situation can be found in most of the small logic circuits. An example is shown in Fig. 5.15 and illustrates a stand-alone 4-to-1 MUX. A stand-alone MUX can be realized as depicted, with the equivalent schematic in Fig. 5.15. The crossbar layout implementation is realized by associating a wire with each branch of the internal stages, whereby a branch is a sequence of transistors connecting the power line to the output buffer stage. The particularity of this multiplexer is that its data inputs

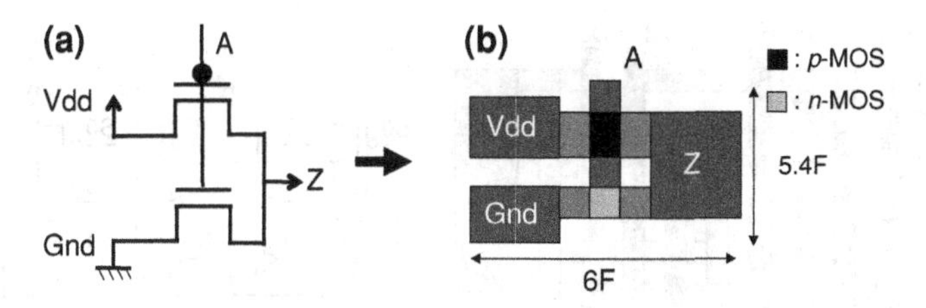

Fig. 5.14 Crossbar inverter structure. **a** Equivalent circuit and **b** layout

control only transistor gates, thus avoiding any signal losses, while the output is buffered to keep the integrity of logic functions independent of the fan-out.

- Alternating Doping Regions

This patterning style consists of alternating *p*- and *n*-type regions, and is particularly suited to build pass-gates between two contact pads. This technique is used in an FPGA-adapted 4-to-1 multiplexer, as depicted in Fig. 5.16. Pass-gates have been placed between the inputs and the output of the MUX. The addressing is realized using poly-Si lines, which control all transistors. Control inputs are complemented by a primary inverter stage.

5.4.4 Performance Evaluation

5.4.4.1 Methodology

In this section, we study the impact of the crossbar technology based on the example of the designed MUX, by evaluating its area, delay and power consumption. This kind of technology can be expected as a near-term solution. In fact, all the technological processes currently exist and only the crossbar organization is fundamentally specific. This means that we can use standard industrial process data for the evaluation. The stand-alone crossbar MUX is thus compared to the equivalent MUX taken from industrial 65 nm design kit. The area is extracted considering the layouts in a 65 nm lithographic node. The delay is extracted from electrical simulation and decomposed into the intrinsic delay and K_{load} factor, which gives the load delay. The model used in the simulation is a scaled FDSOI transistor card for the PSP model. The power consumption is extracted from electrical simulations with a FO4 load. The circuit is operated at the maximum frequency achievable by the gate with the FO4 load. At this frequency, we swept all possible input vector combinations and we averaged the power consumption.

First, we simulated these metrics for the parasitic-free circuit, and then we introduced the parasitics and assessed the designed circuit under different conditions in order to minimize the impact of those parasitics.

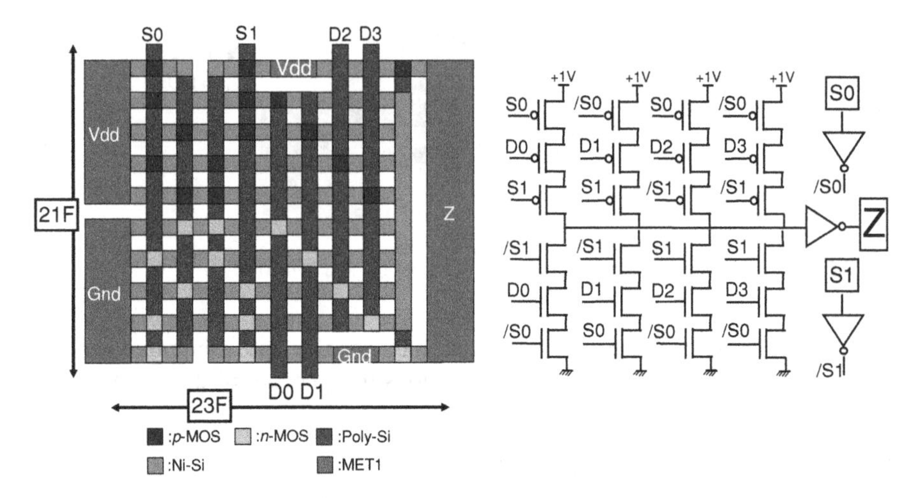

Fig. 5.15 Stand-alone 4:1 MUX crossbar implementation (separate doping regions) and equivalent schematic

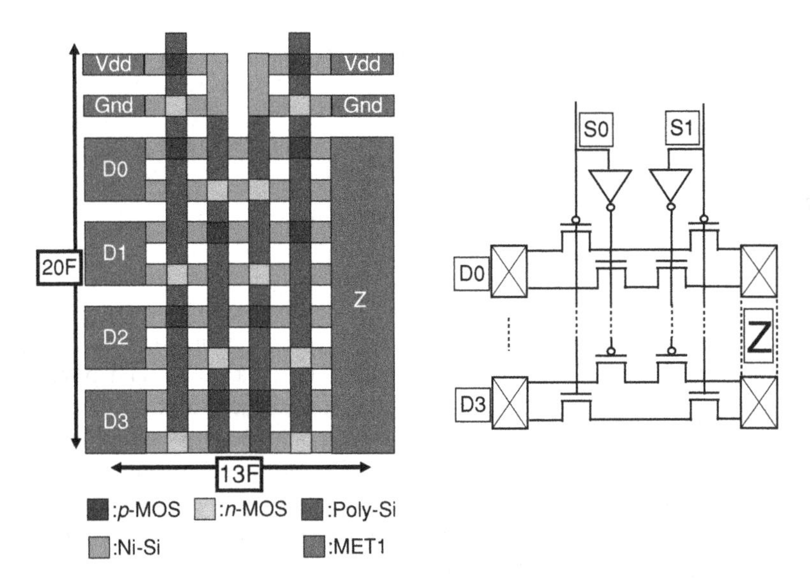

Fig. 5.16 FPGA-suited 4:1 MUX crossbar implementation (alternating doping regions) and equivalent schematic

Due to their very compact layout, a large number of parasitic devices have to be considered in the crossbar layout. A zoom on the parasitics at and around a cross point is shown in Fig. 5.17. We distinguish between two types of cross points: the *active cross points* operating as field-effect transistors, with a poly-Si line electrostatically controlling the Si line underneath it; and the *passive cross points* that are formed by a poly-Si line crossing a passivated highly doped Si line, without

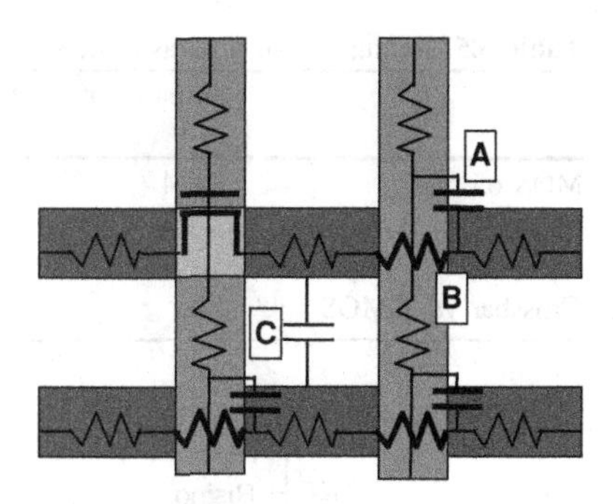

Fig. 5.17 Modeling of parasitic devices

forming any FET. In our process proposal, the passive cross points are formed by a highly-doped conductive Si wire crossing a salicide poly-Si line. The two lines are separated solely by a thin layer of gate dielectric. Thus, the areas under the poly-Si lines, which are not used as a FET channel, result in a high parasitic capacitance between the two wires (see element A in Fig. 5.17). Moreover, we modelled the parasitic resistance through the doped conductive areas (see element B in Fig. 5.17), as well as the electrostatic inter-wire coupling using the capacitive element C in Fig. 5.17.

A parametric study was carried out to study the influence of the following design and technology parameters: *width of horizontal N-type wires* (F_{HN}), *width of horizontal P-type wires* (F_{HP}), *width of vertical poly-Si wires* (F_V), *width of insulator lines between adjacent wires* (F_I). All possible widths are multiples of the lithographic poly-Si half-pitch F.

5.4.4.2 Performance Estimations with Parasitic-Free Circuits

The considered metrics were evaluated on an ideal structure by neglecting all parasitics. The results are shown in Table 5.5.

Such a crossbar structure with its high integration density is able to improve drastically the area by a factor of 6, while the intrinsic delay and the average power are improved respectively by 1.3 and 1.9. While the area gain is due to the compactness of the crossbar structure, the improvement of the other metrics is mainly due to the good isolation of the nanowires in the FDSOI technology compared to bulk. In Table 5.5, we also see that load influence on delay is higher than in bulk. This is due to the confined crossbar dimensions, which constrain transistor drive strength in $W = L = F$ (Sect. 5.4.3.1).

5.4.4.3 Impact of the Width of Horizontal Wires Including Parasitics

Figures 5.18 and 5.19 show the propagation delay in rising and falling edges for varying F_{HP} and F_{HN}, with an inverted output. When F_{HP} (resp. F_{HN}) increases, the falling (resp. rising) propagation delay decreases and tends to a minimum value

Table 5.5 Evaluation of a parasitic-free 4-to-1 MUX in lithographic crossbar technology

	Area (μm^2)	Intrinsic delay (ps)	K_{Load} (ps fF^{-1})	Average power at 650 MHz (μW)
MOS 65 nm	10.4	155.2	11.2	1.47
Crossbar Compact dimensions	1.74	121.8	97.5	0.77
Crossbar vs. CMOS	×6	×1.3	×0.1	×1.9

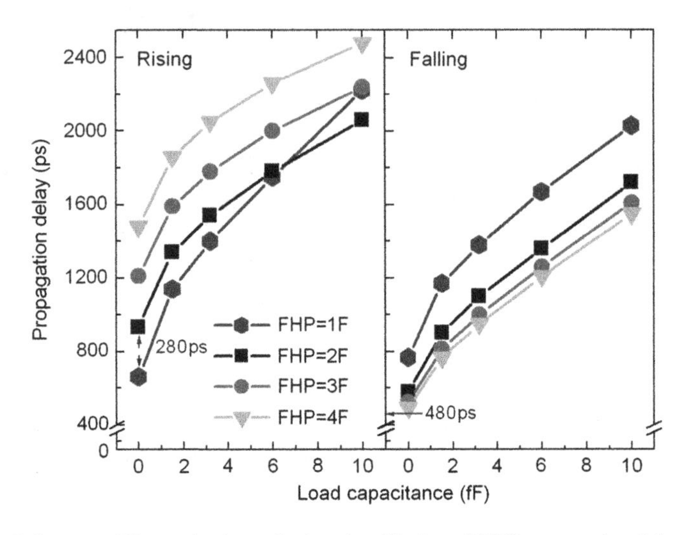

Fig. 5.18 Influence of P-type horizontal wire size (F_{HP}) on MUX propagation delay (output of the inverter) ($F_{HN} = F_V = F_I = F = 60$ nm)

close to 480 ps. When F_{HP} (resp. F_{HN}) increases, the rising intrinsic propagation delay increases linearly by about 280 ps (resp. 400 ps) at each increment. F_{HP} and F_{HN} impact the crossbar structure in several ways.

When F_{HP} (resp. F_{HN}) increases, the width of the *p*-type (resp. *n*-type) transistors increases, but so do also some parasitic devices. While the parasitic resistances (B) are reduced, the parasitic capacitances (A) at the cross points are increased. While increasing F_{HP} (resp. F_{HN}) gives a better response at the output of the inverter stage with respect to the falling (resp. rising) time by improving the transistor drive strength, the rising (resp. falling) response, is however essentially driven by the parasitic capacitance contribution. To minimize this influence factor, we must keep the size as close as possible to the minimum *F*.

5.4.4.4 Impact of the Parasitic Capacitances

Figure 5.20 shows the impact of inter-wire insulator size on the propagation delay. We notice that the propagation delay remains almost constant for any value of F_I.

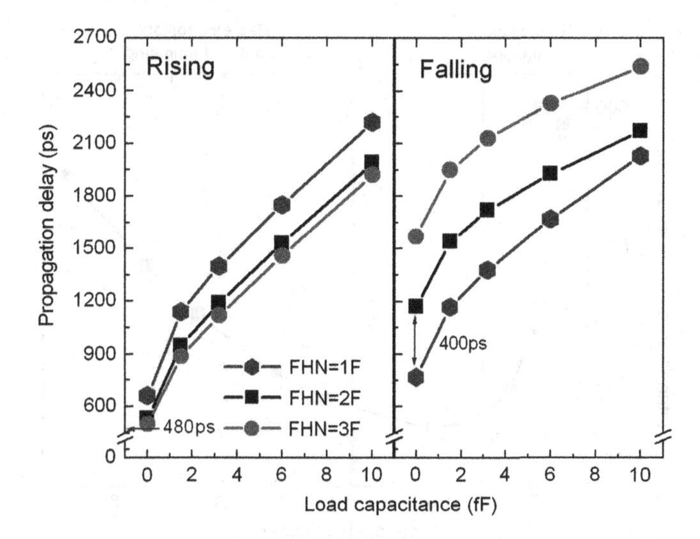

Fig. 5.19 Influence of N-type horizontal wire size (FHN) on MUX propagation delay (output of the inverter) ($F_{HP} = F_V = F_I = F = 60$ nm)

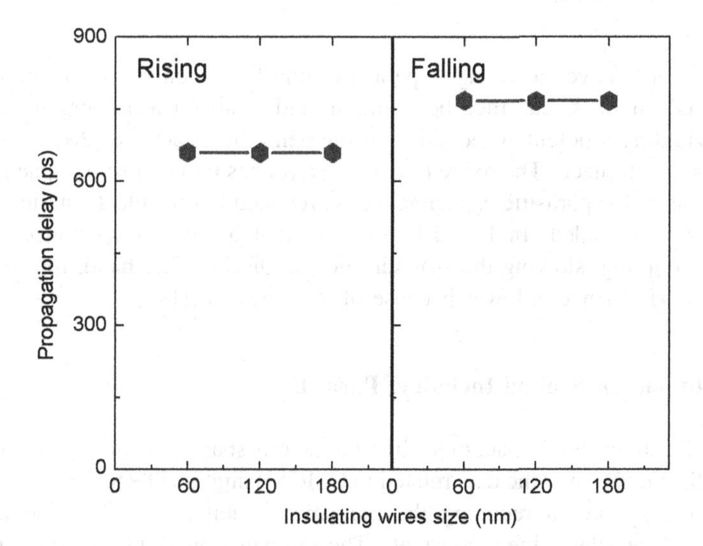

Fig. 5.20 Influence of insulating wire size (F_I) on MUX propagation delay ($F_{HP} = F_{HN} = F_V = F = 60$ nm)

The spacing between the lines is directly related to the crosstalk parasitic capacitances between wires (C). Its influence appears negligible compared to other parasitic capacitances. Nevertheless, this result might not be adapted to aggressive scaling, and an in-depth study will be required.

Figure 5.21 shows the impact of gate oxide thickness (T_{OX}) on the propagation delay. As depicted in this figure, the capacitances formed under the passive cross

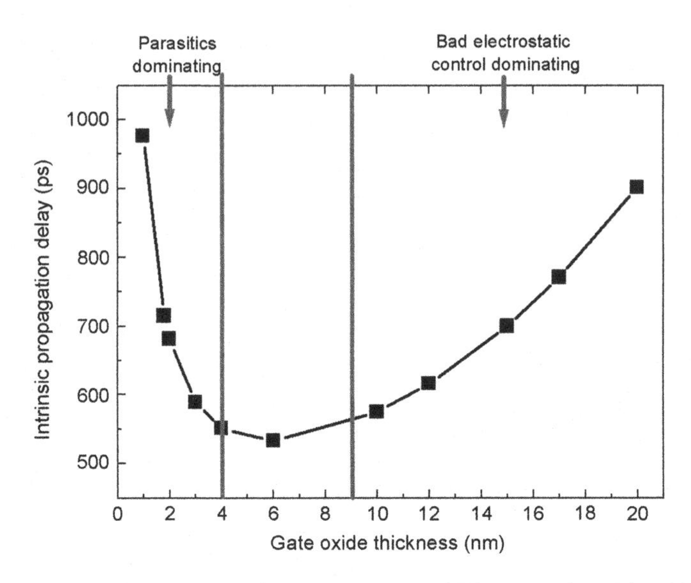

Fig. 5.21 Influence of gate oxide thickness (T_{OX}) on MUX propagation delay ($F_{HP} = F_{HN} = F_V = F_I = F = 60$ nm)

points appear to have the strongest parasitic contribution on the propagation delay. An optimal thickness can then be found around 6 nm for a lithographic pitch of 65 nm, yielding a potential reduction of the delay by a factor of 2×, if the oxide thickness is optimized. The oxide thickness (T_{OX}) has two opposite impacts. While T_{OX} increases, the parasitic capacitances are reduced but the electrostatic effect on the FETs is degraded. In Fig. 5.10, we see that a small T_{OX} induces a large parasitic coupling, slowing the structure down. On the other hand, for very large T_{OX}, the performance is lower because of the slower FETs.

5.4.4.5 Impact of Scaling Including Parasitics

Figure 5.22 shows the impact of scaling on the crossbar implementation compared to CMOS. We obtained the data related to CMOS using the ITRS data [48]. Scaling plays a major role in reducing the parasitic capacitances, while the parasitic resistances are maintained constant. The propagation delay is then reduced accordingly, and thus the performance scales advantageously for crossbar implementation compared to conventional CMOS. For aggressively scaled nodes (<22 nm), the crossbars maintain their advance in terms of area and finally surpass their CMOS counterpart in terms of performance, as shown in Table 5.6. Finally, we found that the average power consumption of crossbars is lower than that of the CMOS implementation by a factor of about 1.5 (1.47 μW vs. 0.99 μW at 650 MHz).

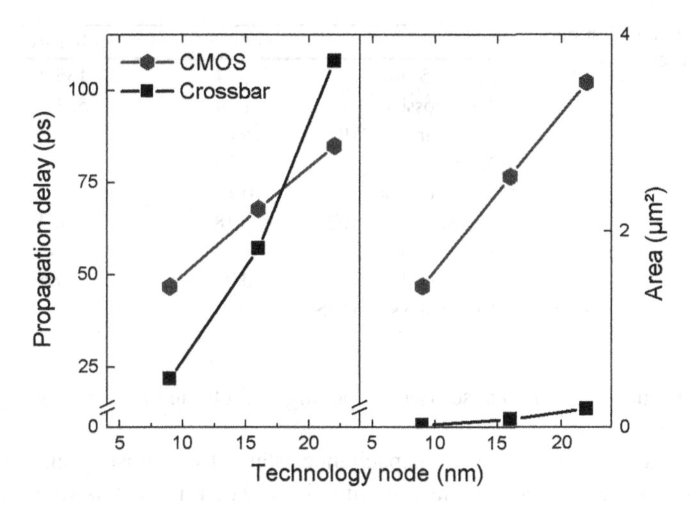

Fig. 5.22 Impact of scaling ($F_{HP} = F_{HN} = F_V = F_I = F$)

5.4.4.6 Performance Estimations Including Parasitics

To summarize, we evaluated the different metrics (area, delay, power), while taking into account the presence of parasitic devices. The evaluated structures have an optimized oxide thickness T_{OX} of 6 nm. The results are shown in Table 5.7. Even if the technology was optimized to minimize the parasitics, we observe a substantial degradation in performance in terms of intrinsic delay and load factor. Nevertheless, the average power consumption of crossbars is lower than that of the CMOS implementation by a factor of about 1.5.

5.4.4.7 Discussion and Guidelines

The previously surveyed simulation results show different design and fabrication scenarios for a crossbar MUX, which can be considered in order to assess the impact of technology on the performance of crossbars in general. The simulations confirm the compactness of the crossbar architecture and its ability to yield up to 6× area saving for the considered MUX design. Moreover, the simulations performed on ideal structures with no capacitive coupling between the different lines also confirm the results reported in the literature [49] on the high performance of the crossbar architecture with respect to CMOS.

However, realistic estimations of the capacitive coupling between the crossbar lines show that their impact cannot be neglected, especially when they are combined with the resistance of the doped semiconducting lines. In this case, performance drops by 4.5×. This result can be understood by the fact that the lines of a crossbar perform the interconnect task. The insulator is intended to mutually shield the parallel lines (lying on the same level) and the perpendicular lines (lying

Table 5.6 Evaluation of scaling with parasitics

	Area (μm^2)	Intrinsic delay (ps)
MOS 65 nm	10.4	155.2
NW crossbar	1.74	533
Crossbar vs. CMOS	×6	×0.29
MOS 22 nm	3.52	84.9
NW crossbar	0.19	107.9
Crossbar vs. CMOS	×18	×0.78
MOS 9 nm	1.44	46.8
NW crossbar	0.02	21.86
Crossbar vs. CMOS	×67	×2.14

on top of each other). These two opposing effects lead to a high parasitic contribution.

The largest impact of parasitic capacitances stems from those lying at the cross points, i.e., those formed by the crossing lines. Their impact is dominant with respect to that of the lines lying between the parallel lines in every layer. In order to address the issue of capacitive coupling between the crossing lines, a solution can be driven from the technology side. While the insulating material between the top and bottom lines has to be optimized by using a thick or a low-K material, this insulator represents at the same time the gate oxide, which should be thin or a high-K material in order to improve the gate control over the channel. These conditions are not compatible, which is a natural result of the double nature of the crossbar: it performs both interconnect and logic with the same resources. Consequently, there is a trade-off between the speed of logic and the speed of interconnects. It is therefore necessary to accept a certain compromise in terms of delay when the insulating material between the crossing lines is chosen. Our study shows that for a SiO_2 gate oxide, the trade-off is optimized for an oxide thickness of about 6 nm.

The investigation we carried out in this section is based on a MUX aiming to illustrate the general trend of crossbars. Because of the double role of this architecture (interleaved interconnect and logic), a trade-off in performance must be found. The technology must optimize the insulating oxide between the crossing layers in order to meet the optimal trade-off. It is worth noticing that the results of these investigations are valid under our technological assumptions of a lithographic and manufacturable process.

5.5 Global Comparisons

In this chapter, we investigated the use of three different technologies to build a compact logic computation cell. Among these three technologies, we can differentiate the first technology, which aim to increase the functionality of the elementary device such as the double-gate carbon nanotube field effect transistor, from the other technologies, which help to increase the density of elementary transistors. In the latter category, we examined a highly speculative approach,

Table 5.7 Evaluation of lithographic crossbar-based structure with parasitics

	Area (μm^2)	Intrinsic delay (ps)	K_{Load} (ps fF^{-1})	Average power at 650 MHz (μW)
MOS 65 nm	10.4	155.2	11.2	1.47
Crossbar compact dimensions	1.74	533	208	0.99
Crossbar vs. CMOS	×6	×0.29	×0.05	×1.48

Table 5.8 Global comparison of analyzed disruptive technology cells

	Middle-term proposals			Short-term proposals	
	CMOS MUX 22-nm	DG-CNTFET cell	Sublithographic crossbar	CMOS MUX 65-nm	Lithographic crossbar
Area (μm^2)	1.19	0.39	0.29	10.4	1.74
Intrinsic delay (ps)	79	149	17	155.2	533
K_{Load} (ps fF^{-1})	5.7	124.5	15.1	11.2	208
Average power (μW)	4	1.78	6.7	1.47	0.99

based on the sublithographic arrangement of nanowires and a more realistic one, which is derived from an industrial FDSOI process flow.

Table 5.8 summarizes the metrics used to evaluate the different proposals. To compare the different technologies objectively, we will split the evaluation of future middle-term solutions (carbon electronics and sublithographic arrangements of nanowires) from that of short-term, potentially immediately manufacturable solutions (lithographic crossbar arrangement). The future solutions are compared to a scaled 22 nm node, while the short-term proposal, based on a standard industrial process, is compared with respect to a 65 nm CMOS node.

It is worth noticing that the technologies improving the density are highly beneficial for area improvement.

Indeed, the best gain is obtained for a lithographic crossbar with a grain of 6×, compared to current CMOS technologies. Regarding the middle-term solutions, the sublithographic crossbar is also the most compact solution.

The best gain in power consumption was reached for carbon electronics, with a gain of about 2× compared to its CMOS counterpart. It appears clear that the good electron transport in carbon structures is at the origin of this.

Regarding the delay performances, the CMOS circuits are still good candidates for future generations of circuits. Nevertheless, performance optimization for the other technologies has not been carried out for this first order evaluation, and performance can certainly be improved after correct sizing and process tuning.

5.6 Conclusion

In this chapter, we have studied how emerging technologies could be used to reduce the size of the computation node for FPGAs. In fact, it is worth noticing that only 14% of the FPGAs area is occupied by the logic blocks. More than 80% is used just by peripheral circuitries. Thus, the FPGA architectural scheme seems quite inefficient if we compare the circuit area used for computation to that of the others. It is thus of high interest to propose new elementary logic circuits, that may lead to alternatives to the standard FPGA scheme.

Emerging technologies have been used to reduce the size of the computation node compared to CMOS. We have surveyed two main ways to do so: *functionality*-improvement based devices and *density*-improvement based devices.

For functionality-based improvements, we surveyed the use of the DG-CNFET technology. Such a device can be configured to *n*-, *p*- or *off*-type depending on the voltage applied to the back-gate terminal of the device. We first presented an integration process for the technology and performed device optimization, in order to ensure correct hypotheses. In-field reconfigurable devices lead to compact reconfigurable logic cells. We performed the evaluation of a previously designed circuit and compared it to scaled CMOS, and showed that the reduction of the number of transistors yields a reduction of 3.1× in area. Power consumption is reduced by 2×, due to the electronic properties of carbon. Finally, the use of back-gate control is generalized to standard dynamic logic cells. We especially used the off-state to merge the clocking transistors in the function paths. Thus, we demonstrate a gain in terms of transistors of up to 50% for the XOR gate.

For the density-based improvement techniques, we surveyed the use of dense crossbar implementations of devices. The crossbar implementation expects to merge connections and active logic in a very small area. Various assumptions for crossbar implementations could be used. Firstly, we envisage the use of sublithographic techniques to realize dense arrays of nanowires that formed FET at their cross points. Sublithographic techniques are based on very speculative bottom-up arrangements of wires at the nanoscale. After describing all technological assumptions, we derived a NASIC approach to build a multiplexer intended to work and to be connected at the lithographic scale. We showed that this technology yields an improvement of 4.1× in area and 4.6× in delay, thanks to the drastic increase of density and the consequent reduction of parasitics.

Sublithographic integration processes are highly speculative and only long-term solutions. The crossbar concept is of course not limited to such an advanced technology and could be extended to lithographic processes. In this way, we derived a FDSOI process flow in order to realize a dense crossbar arrangement of transistors, with lithographic dimensions. We then proposed a layout methodology for standard circuits. After describing all the parasitic elements in the structure, we performed a circuit-driven optimization of the technology to find the best trade-off between the parasitics and the transistor performances. We showed that this solution yields an improvement of 6× in area and 1.48× in power consumption.

This is due to the crossbar arrangement and the performance of FDSOI technology in terms of power. We also showed that this approach could be expected to outperform the standard CMOS scheme from the 16 nm technological node onward.

To conclude, various technologies have been used with the same objective: build a compact and improved logic cell. In the light of these results, we can clearly see that a trade-off must be found depending on the required specifications in terms of delay, area or power. Nevertheless, we should notice that all the cells reduce the area figures. This may lead to the use of a new highly compact elementary logic cell for high-performance computation. This basic block will be envisaged as a new seed for architectural organization, discussed in the next chapter.

References

1. M. Lin, A. El Gamal, Y.-C. Lu, S. Wong, Performance benefits of monolithically stacked 3-D FPGA. IEEE Trans. Comput. Aided Des. Integr. Circuits Syst. **26**(2), 216–229 (2007)
2. A. DeHon, M. J. Wilson, Nanowire-based sublithographic programmable logic arrays, in *Proceedings of the 2004 ACM/SIGDA 12th international Symposium on Field Programmable Gate Arrays* (2004)
3. T. Wang, P. Narayanan, C.A. Moritz, Combining 2-level logic families in grid-based nanoscale fabrics, in *IEEE/ACM International Symposium on Nanoscale Architectures (NANOARCH)*, October 2007
4. S. Iijima, T. Ichihashi, Single-shell carbon nanotubes of 1 nm diameter. Nature **363**, 603–605 (1993)
5. K.S. Novoselov, A.K. Geim, S.V. Morozov, D. Jiang, Y. Zhang, S.V. Dubonos, I.V. Grigorieva, A.A. Firsov, Electric field effect in atomically thin carbon films. Science **306**(5696), 666–669 (2004)
6. M.S. Dresselhaus, G. Dresselhaus, P. Avouris, *Carbon Nanotubes: Synthesis, Structure, Properties, and Applications* (Springer, Berlin, 2001)
7. Z. Yao, C.L. Kane, C. Dekker, High-field electrical transport in single-wall carbon nanotubes. Phys. Rev. Lett. **84**, 2941–2944 (2000)
8. M.S. Fuhrer, B.M. Kim, T. Duerkop, T. Brintlinger, High-mobility nanotube transistor memory. Nano Lett. **2**, 755–759 (2002)
9. A.H. Castro Neto, F. Guinea, N.M.R. Peres, K.S. Novoselov, A.K. Geim, The electronic properties of grapheme. Rev. Mod. Phys. **81**(1), 109–163 (2009)
10. I. Meric, M. Han, A.F. Young, B. Ozyilmaz, P. Kim, K.L. Shepard, Current saturation in zero-bandgap, top-gated grapheme field-effect transistors. Nat. Nanotechnol. **3**(11), 654–659 (2008)
11. Y.-M. Lin, K.A. Jenkins, A. Valdes-Garcia, J.P. Small, D.B. Farmer, P. Avouris, Operation of grapheme transistors at gigahertz frequencies. Nano Lett. **9**(1), 422–426 (2009)
12. I. Meric, N. Baklitskaya, P. Kim, K.L. Shepard, RF performance of top-gated, zero-bandgap graphene field-effect transistors, in *IEDM Technical Digest* (2008)
13. J.S. Moon, D. Curtis, M. Hu, D. Wong, C. McGuire, P.M. Campbell, G. Jernigan, J.L. Tedesco, B. VanMil, R. Myers-Ward, C. Eddy Jr, D.K. Gaskill, Epitaxial-graphene RF field-effect transistors on Si-face 6H-SiC substrates. IEEE Electron Dev. Lett. **30**(6), 650–652 (2008)

14. C. Berger, Z. Song, T. Li, X. Li, A.Y. Ogbazghi, R. Feng, Z. Dai, A.N. Marchenkov, E.H. Conrad, P.N. First, W.A. de Heer, Ultrathin epitaxial graphite: 2D electron gas properties and a route toward graphene-based nanoelectronics. J. Phys. Chem. B **108**(52), 19912–19916 (2004)
15. X. Li, X. Wang, L. Zhang, S. Lee, H. Dai, Chemically derived, ultrasmooth graphene nanoribbon semiconductors. Science **319**(5867), 1229–1232 (2008)
16. K.-T. Lam, G. Liang, An ab initio study on energy gap of bilayer graphene nanoribbons with armchair edges. Appl. Phys. Lett. **92**(22), 223106 (2008)
17. C.L. Lu, C.P. Chang, Y.C. Huang, J.M. Lu, C.C. Hwang, M.F. Lin, Low-energy electronic properties of the ab-stacked few-layer graphites. J. Phys. Condens. Matter **18**(26), 5849 (2006)
18. H. Min, B. Sahu, S.K. Banerjee, A.H. MacDonald, Ab initio theory of gate induced gaps in graphene bilayers. Phys. Rev. B **75**(15), 155115 (2007)
19. Y. Zhang, T.-T. Tang, C. Girit, Z. Hao, M.C. Martin, A. Zettl, M.F. Crommie, Y.R. Shen, F. Wang, Direct observation of a widely tunable bandgap in bilayer graphene. Nature **459**(7248), 820 (2009)
20. R. Sordan, K. Balasubramanian, M. Burghard, K. Kern, Exclusive-OR gate with a single carbon nanotube. Appl. Phys. Lett. **88**(5), 053119–0531193 (2006)
21. Y.-M. Lin, J. Appenzeller, J. Knoch, P. Avouris, High-performance carbon nanotube field-effect transistor with tunable polarities. IEEE Trans. Nanotechnol. **4**(5), 481–489 (2005)
22. J. Appenzeller, J. Knoch, V. Derycke, R. Martel, S. Wind, P. Avouris, Field-modulated carrier transport in carbon nanotube transistors. Phys. Rev. Lett. **89**(12), 126801–126804 (2002)
23. A.A. Kane, T. Sheps, E.T. Branigan, V.A. Apkarian, M.H. Cheng, J.C. Hemminger, S.R. Hunt, P.G. Collins, Graphitic electrical contacts to metallic single-walled carbon nanotubes using Pt electrodes. Nano Lett. **9**(10), 3586–3591 (2009)
24. N. Patil, A. Lin, E.R. Myers, K. Ryu, A. Badmaev, C. Zhou, H.-S.P. Wong, S. Mitra, Wafer-scale growth and transfer of aligned single-walled carbon nanotubes. IEEE Trans. Nanotechnol. **8**(4), 498–504 (2009)
25. I. O'Connor, J. Liu, F. Gaffiot, F. Pregaldiny, C. Lallement, C. Maneux, J. Goguet, S. Fregonese, T. Zimmer, L. Anghel, T.-T. Dang, R. Leveugle, CNTFET modeling and reconfigurable logic-circuit design. IEEE Trans. Circuits Syst. I Regul. Pap. **54**(11), 2365–2379 (2007)
26. N.F. Goncalves, H. De Man, NORA: a racefree dynamic CMOS technique for pipelined logic structures. IEEE J. Solid State Circuits **18**(3), 261–266 (1983)
27. K.H. Yeo, S.D. Suk, M. Li, Y.-Y. Yeoh, K.H. Cho, K.-H. Hong et al., Gate-all-around (GAA) twin silicon nanowire MOSFET (TSNWFET) with 15 nm length gate and 4 nm radius nanowires, in *International Electron Devices Meeting 2006*, 11–13 Dec 2006, pp. 1–4
28. Y. Huang, X. Duan, Y. Cui, L. Lauhon, K. Kim, C.M. Lieber, Logic gates and computation from assembled nanowire building blocks. Science **294**, 1313–1317 (2001)
29. Y. Wu, J. Xiang, C. Yang, W. Lu, C.M. Lieber, Single-crystal metallic nanowires and metal/semiconductor nanowire heterostructures. Nature **430**, 61–65 (2004)
30. A. DeHon, Array-based architecture for FET-based, nanoscale electronics. IEEE Trans. Nanotechnol. **2**(1), 23–32 (2003)
31. T. Wang, P. Narayanan, M. Leuchtenburg, C.A. Moritz, NASICs: a nanoscale fabric for nanoscale microprocessors, in *IEEE International Nanoelectronics Conference*, 24–27 Mar 2008
32. R.S. Wagner, W.C. Ellis, Vapour–liquid–solid mechanism of single crystal growth. Appl. Phys. Lett. **4**(5), 89–90 (1964)
33. Y. Cui, L.J. Lauhon, M.S. Gudiksen, J. Wang, C.M. Lieber, Diameter-controlled synthesis of single crystal silicon nanowires. Appl. Phys. Lett. **78**(15), 2214–2216 (2001)
34. Y. Wu, Y. Cui, L. Huynh, C.J. Barrelet, D.C. Bell, C.M. Lieber, Controlled growth and structures of molecular-scale silicon nanowires. Nanoletters **4**(3), 433–436 (2004)

35. C.J. Kim, D. Lee, H.S. Lee, G. Lee, G.S. Kim, M.H. Jo, Vertically aligned Si intrananowire p-n diodes by large-area epitaxial growth. Appl. Phys. Lett. **94**, 173105 (2009)
36. L.J. Lauhon, M.S. Gudiksen, D. Wang, C.M. Lieber, Epitaxial core-shell and core-multi-shell nanowire heterostructures. Nature **420**, 57–61 (2002)
37. A. Ulman, *An Introduction to Ultrathin Organic Films: From Langmuir-Blodgett to Self Assembly* (Academic Press, New York, 1991)
38. R.K. Brayton, A.L. Sangiovanni-Vincentelli, C.T. McMullen, G.D. Hachtel, *Logic Minimization Algorithms for VLSI Synthesis* (Kluwer Academic Publishers, New York, 1984)
39. T. Wang, M. Ben-Naser, Y. Guo, C.A. Moritz, *Wire-Streaming Processors on 2-D Nanowire Fabrics* (Nano Science and Technology Institute, Santa Clara, 2005)
40. B. Cousin, M. Reyboz, O. Rozeau, M. Jaud, T. Ernst, J. Jomaah, A continuous compact model of short-channel effects for undoped cylindrical gate-all-around MOSFETs, in *9th Workshop on Compact Modeling* (2010), pp. 793–796
41. B. Cousin, M. Reyboz, O. Rozeau, M.-A. Jaud, T. Ernst, J. Jomaah, A unified short-channel compact model for cylindrical surrounding-gate MOSFET. Solid State Electronics **56**(1), 40–46 (2011)
42. Y. Chen, G.-Y. Jung, D.A. Ohlberg, X. Li, D.R. Stewart, J.O. Jeppesen et al., Nanoscale molecular-switch crossbar circuits. Nanotechnology, **14**(4), 462–468 (2003)
43. D. Whang, S. Jin, Y. Wu, C.M. Lieber, Large-scale hierarchical organization of nanowire arrays for integrated nanosystems. Nano Lett. **3**(9), 1255–1259 (2003)
44. D.B. Strukov, K.K. Likharev, CMOL FPGA: a reconfigurable architecture for hybrid digital circuits with two-terminal nanodevices. Nanotechnology, **16**(6), 888–900 (2005)
45. T. Ernst, L. Duraffourg, C. Dupré, E. Bernard, P. Andreucci, S. Bécu et al., Novel Si-based nanowire devices: will they serve ultimate MOSFETs scaling or ultimate hybrid integration?, in *IEEE International Electron Devices Meeting* (2008)
46. C. Dupré, A. Hubert, S. Becu, M. Jublot, V. Maffini-Alvaro, C. Vizioz et al., 15 nm-diameter 3D stacked nanowires with independent gates operation: ΦFET, in *IEEE International Electron Devices Meeting*, 15–17 Dec 2008
47. D. Lenoble, A. Grouillet, The fabrication of advanced transistors with plasma doping. Surf. Coat. Technol. **156**, 262–266 (2002)
48. Executive Summary, Updated Edition, International Technology Roadmap for Semiconductors (2010), http://www.itrs.net/Links/2010ITRS/Home2010.htm
49. A. DeHon, Architecture approaching the atomic scale, in *Proceedings of the 33rd European Solid-State Circuits Conference*, ESSCIRC (2007)

Chapter 6
Disruptive Architectural Proposals and Performance Analysis

Abstract In this chapter, we explore disruptive architecture proposals. In the previous chapter, we showed that it is possible to obtain very compact reconfigurable in-field computation cells. Since these cells require architectural modifications, we proposed an architecture for this compact logic, characterized by the association of a logic layer, to adapt the granularity and the use of fixed interconnection topologies to reduce the routing impact. To compare this approach with conventional FPGAs in an objective way, it was necessary to develop a specific toolflow suited to our requirements, able to describe the designed architecture. Based on the VTR toolflow, the tool integrates fixed topology routing and the specific organization of the layered architecture. Benchmarking simulations were performed. In a first approach, a local exploration of the proposed layer is done, in order to study the impact of the fixed interconnect topologies. We showed that the Modified Omega topology gives the best mapping rates on the structure with about 90% of mapping success for 6-node graphs. In a second approach, complete architectural benchmarking was conducted and we showed that the proposed architecture leads to an improvement, in area saving, of 46% in average, with respect to CMOS. We also discovered that the routing delay is less distributed and tends to be more controllable than in the traditional approach.

In the previous chapter, we introduced compact logic cells. With emerging technologies, an elementary logic function can be efficiently implemented by hardware. The root of this higher efficiency can be twofold: firstly, the augmented functionality of the device allows the number of transistors required per function to be reduced, and secondly, higher device integration density enables more devices to be embedded in a given area, which obviously shrinks the size of the implemented logic. However, this aspect does not at first appear to be a particular advantage for reconfigurable logic. In Chap. 2, we stated that only a low proportion of the overall circuit is used by elementary combinational and sequential logic (14% of total area), as compared to routing and memories (86% of total

P.-E. Gaillardon et al., *Disruptive Logic Architectures and Technologies*,
DOI: 10.1007/978-1-4614-3058-2_6,

area). Nevertheless, with the emergence of a novel ultra-compact logic cell, we face new architectural requirements, which could lead to a change in balance between logic resources and routing/memory resources. In this chapter, we will describe a new architecture for reconfigurable ultra-fine grain computing. To address the specificity of this new architecture, we will adapt the benchmarking methodology and present a specific packing tool. Finally, we will size the new architectural model and compare it to a standard MOS counterpart.

6.1 Introduction

In Chap. 2, we already presented the *Field Programmable Gate Array* (FPGAs) architecture. This is currently the most common architectural scheme for reconfigurable logic. The architecture is designed for current CMOS technologies, and its regularity is compatible with most scaled technological nodes. The logic is performed through the use of *Configurable Logic Blocks* (CLBs), able to perform coarse-grain combinational and sequential logic functions. These CLBs are arranged into islands and interconnected through a complete (but resource-limited) programmable interconnect. The CLBs are formed by several *Basic Logic Elements* (BLEs) interconnected by a fully programmable interconnect infrastructure. BLEs are composed of *Look-Up Table* (LUT) and *Flip-Flop* (FF) elements. In modern FPGAs, several other blocks can be found, in order to perform specific optimized tasks, such as hardware multipliers or memory banks. This heterogeneous nature of the structures helps to optimize the performance for specific application classes. Nevertheless, since we focus on generic reconfigurable logic structures, we will consider in this chapter only homogeneous structures. Obviously, in conventional CMOS FPGAs, all these building blocks are sized to ensure a correct granularity of the structure and to maximize the performance of the assembled system.

In this book, we looked at disruptive technologies and saw that they could lead to very compact logic elements. These elements will be proposed as the foundation stone for all reconfigurable arrangements. Nevertheless, the organization of such blocks induces new problems to solve. As an illustration, if we consider the in-field reconfigurable cell proposed in Sect. 5.2.4, the compactness of the cell means that it is not possible to transpose directly the traditional FPGA architecture to our ultra-fine grain cell, since complete interconnectivity would be required between the elements. Again, we may consider that about 43% of silicon area is used for interconnects, while only 14% is used for logic [1]. However, in the case of ultra fine-grain logic, this ratio would further worsen, since the routing structures do not scale in the same way as logic. Thus, there will be a significant area imbalance between logic and interconnect, consequently leading to a severe loss of efficiency in the structure. Another area imbalance question might be raised by memories. Indeed, even if the size of the logic element has been decreased, the amount of memory required remains the same. In Chap. 3, proposals have been made to reduce the complexity of memory

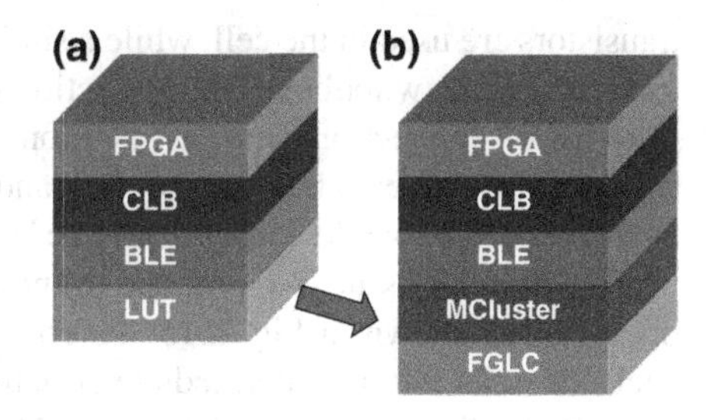

Fig. 6.1 FPGA model (**a**) and fine-grain modified levels (**b**)

in an FPGA. Nevertheless, their area impact is not zero. It is thus fundamental to bear in mind the extra circuitry required by these configuration memories during performance evaluation. Finally, another constraint is present that of reliability. The use of billions of unreliable emerging devices can be handled by the use of specific fault-tolerance mechanisms. Even if these questions are not addressed in this thesis, it is important to keep this aspect in mind.

To obtain a credible disruptive block and architecture based on emerging technologies, all these considerations must be addressed. In this chapter, we propose an architectural scheme that addresses these issues. We will specifically address the questions regarding the use of ultra-fine grain logic cells in a reconfigurable circuit. It is worth noticing that the resulting architecture will be in aligned with the architectural template of the study. Nevertheless, due to the specificity of the ultra-fine grain approach, the template will be enhanced. Then, in order to evaluate in an objective way the potential of the structures and technologies with respect to their CMOS counterparts, we will also propose specific tools. This tool flow will be compatible with standard approaches, and will only be enhanced to address the specific architecture. Finally, we will evaluate the proposal and compare it to a standard equivalent circuit.

6.2 Architectural Proposals

6.2.1 Introduction

As already mentioned, FPGAs possess a layered structure, as depicted in Fig. 6.1a. Four layers are traditionally considered: elementary logic with *Look-Up Tables* (LUT layer); fine grain combinational and sequential logic with *Basic Logic Elements* (BLE layer); coarse grain logic with *Configurable Logic Blocks* (CLB layer); and finally an island style arrangement (FPGA layer). All blocks were detailed in Chap. 2.

In our approach, we replace the traditional LUTs by a *Fine Grain Logic Cell* (FGLC). However, if we were to project this onto the conventional approach, each FGLC would be directly connected to a full connectivity unit, such as a set of switchboxes. Considering the in-field reconfigurable gate as an example, only seven transistors are used in the cell, while a similar number of transistors (at least six) are used for a 1-bit switchbox. This projection would therefore result in a large overhead in terms of connections and, as mentioned previously, would worsen the already significant imbalance between routing and logic resources at the FPGA level. The overall structure would consequently be left with an intrinsic granularity imbalance.

To solve this issue, we propose to modify the FPGA scheme, with the layered organization shown in Fig. 6.1b. To correct the granularity issues, we propose to pack the cells into an intermediate structure which is compatible in terms of size with the traditional levels of FPGAs: *MClusters* (for Matrix Clusters). Thus, we obtain fine-grain logic blocks with enhanced functionality with respect to conventional LUTs. Within this organization, the connectivity issue between logic cells must be addressed, considering that full connectivity is not an option. We thus propose a solution based on fixed interconnect topologies. Finally, the convergence with the conventional FPGA scheme will occur at the BLE level. In particular, the MClusters will be used as the replacement block for LUTs.

6.2.2 MCluster Organization

6.2.2.1 MCluster: Adding a "Logic Layer"

At the MCluster layer, elementary ultra-fine grain logic cells are used, where the term "ultra-fine grain" is twofold. First, it covers the size of the cell, which is highly compact, as opposed to its CMOS equivalent. Second, it also covers the granularity of the logic functions covered. In an FPGA, the smallest block is in general the LUT, and it has been shown that an LUT with four-inputs is the best solution for routability efficiency [2]. With a four-input LUT, it is possible to realize 65,536 (2^{2^4}) functions. The targeted logic block (the in-field reconfigurable logic cell) uses a more restricted set of functions. This means that we also introduce an imbalance between fine-grain logic and ultra-fine grain logic.

To increase the logic coverage of the structure, we propose a 2D matrix assembly of ultra-fine grain logic. The logic cells are arranged in a layered structure, with connections only existing between adjacent layers in order to avoid long connections and to maximize local connectivity. This organization, called *MCluster,* is depicted in Fig. 6.2 and will handle fine-grain logic operations in the proposed organization. In the following, MClusters will be defined by the w (width) and d (depth) parameters and written as: MCluster_d_w.

6.2.2.2 MCluster: Simplifying the Interconnect Overhead

For intra-matrix interconnect, any total interconnectivity topology is prohibited, because of the associated wiring complexity and the extra area requirements. Instead, and through analogy to computer networks, our approach is to adapt

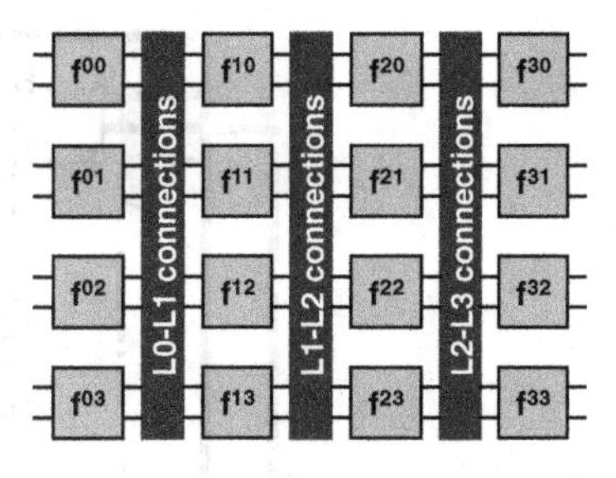

Fig. 6.2 MCluster approach for reconfigurable architectures (MCluster_4_4)

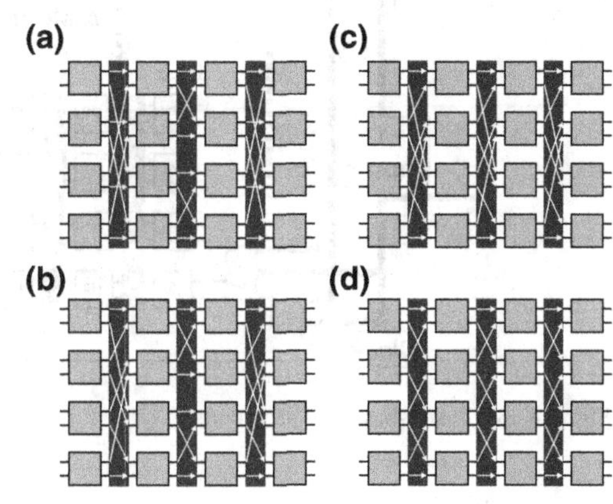

Fig. 6.3 Matrix of 16 reconfigurable gates with fixed interconnect topology. **a** Banyan, **b** Baseline, **c** Flip, **d** Modified Omega)

incomplete interconnection sets to the matrix architecture. In fact, *Multistage Interconnection Networks* (MINs) are designed to interconnect computing layers in an efficient way and can be applied in this context. In computer science, MINs are used to interconnect layers of switchboxes in order to route information packets only. This concept has been reported several times in interconnect strategies for VLSI, for example in [3, 4], and used in FPGAs to reduce the complexity of wiring through logic blocks [5, 6], by proposing an architectural organization for the routing circuits. It is worth noticing that the main difference here, with respect to the network context, is that switchboxes have been replaced by logic cells, thus introducing computing directly inside the network. Furthermore, the use of very local MIN-style interconnect has a drastic impact on the size and the wire length is reduced accordingly.

There are many MIN topologies and combinations, but in this work, we focus on four typical permutations [7, 8]: Banyan (Fig. 6.3a), Baseline (Fig. 6.3b), Flip (Fig. 6.3c) and Modified Omega (Fig. 6.3d), where the modifications to standard Omega maximize the shuffling in this topology. Since the interconnect topology is fixed and static, the choice of topology must be made by the designer during architectural development. We will discuss the intrinsic performance metrics of each topology in the following sections of this chapter.

While fixed interconnect topologies are useful for area considerations, we must also consider the cost in terms of logic mapping performance. In this context, we could explore trade-offs between fixed interconnection patterns and full

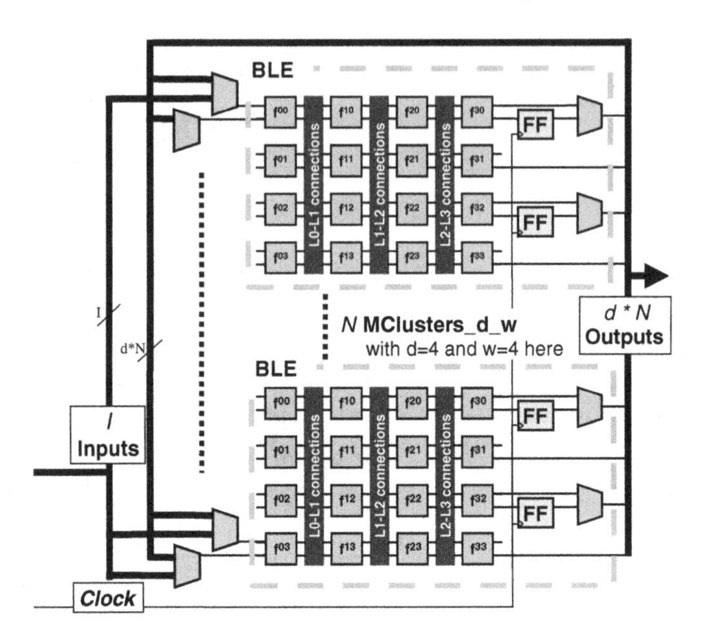

Fig. 6.4 MCluster-based CLB proposal

connectivity. In particular, since we have seen that vertical nanowire field-effect transistor technologies are useful to build smart vias, we could envisage a topology, which can be reconfigured between two or more sets of interconnect, with the number of interconnect sets quite low to avoid a large number of configuration signals.

6.2.3 BLE and CLB Organization

We described above a matrix arrangement for ultra-fine grain logic cells. This arrangement is able to perform combinational logic functions and can thus be considered to be a replacement element at the LUT level. We now consider how the BLE/CLB organization should be adapted using the MClusters.

Figure 6.4 shows the organization of the logic blocks at the CLB level. The BLE are composed of *MClusters_d_w* used to perform the combinational logic. In a traditional approach, all outputs of the LUTs may be latched and multiplexed to the CLB outputs. In our approach, a partial depopulation of the output latches is implemented as shown in Fig. 6.4, where one of the two outputs are not latched. The subsequent CLB organization remains similar to the LUT approach, where all outputs of the BLEs are connected to the global reconfigurable interconnect, and are also fed back to the inputs of the CLB, thus achieving a full connectivity pattern. The *d*N* feedback signals and the *I* inputs of the CLB may be routed to any of the MClusters by using the reconfigurable multiplexers. Finally, a single clock signal is used to control the sequential elements.

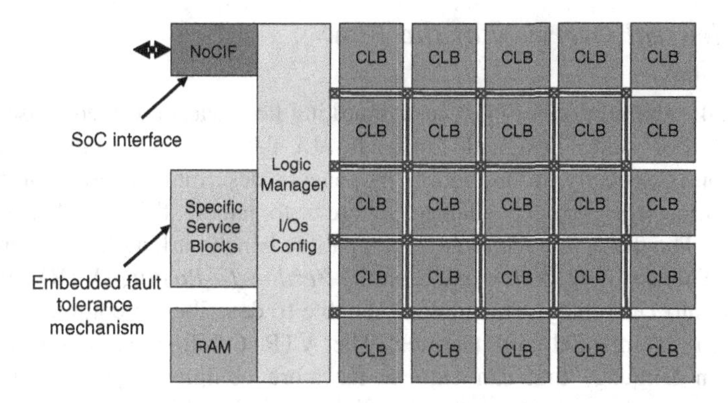

Fig. 6.5 Final FPGA layer organization

6.2.4 FPGA Final Organization

At the top level of organization, the granularity (as well as the configuration and routing) assessments are the same as for conventional FPGAs. Thus, we will consider the FPGA layer, as depicted in Fig. 6.5. Hence, the logic blocks are interconnected by a complete interconnect set, but limited in terms of resources (i.e. all the connections can not be realized simultaneously). As in conventional FPGAs, this architecture uses switchboxes and connection boxes to ensure the connectivity between the CLBs. Nevertheless, we can see from Fig. 6.5 that this "sea of logic" is not directly connected to I/O blocks. While the standard FPGA approach could be used, the reconfigurable circuit is connected in this approach to a logic manager, which can then communicate data to global interconnect through a Network-On-Chip interface. In this way, the architecture can be integrated for example as a reconfigurable hardware accelerator in a Multi-Processor System-on-Chip or in a heterogeneous circuit. Finally, specific blocks are also included, such as a fault-tolerant logic router and a performance improvement decision controller. This kind of IP can operate as a service block to help in re-defining the structural placement and routing, allowing defective CLBs to be avoided, or power consumption to be optimized in real-time throughout the circuit.

6.3 Benchmarking Tool

In the previous section, a specific organization for ultra-fine grain logic cells was described. To analyze performance gain from an application point of view and compare it objectively to existing approaches, we must evaluate the performance of its implementation of well-known circuits (benchmarks). Thus, we propose a complete benchmarking tool flow suited to our proposal, as well as to existing architectures.

6.3.1 General Overview of the Flow

In Chap. 4, the traditional FPGA benchmarking flow was described. Based on the *VPR* 5.0 tool suite, designed to handle LUT-based logic, the architectural description is basically limited to a homogeneous description of logic blocks. The toolflow must therefore be adapted in order to handle the complexity of our structures. The disruptive technology compatible benchmarking flow is depicted in Fig. 6.6. This toolflow is now based on the *Verilog-To-Routing* (VTR) [9] project from Toronto University. This toolflow is able to describe the complexity present in modern commercial architectures. The VTR 6.0 flow uses a logic block description language that can express far more complex logic blocks than is currently possible with any other publicly available toolsets. It can describe complex logic blocks with arbitrary internal routing structures, can allow arbitrary levels of hierarchy within the logic block and can assign different functional modes to blocks. The latter property is necessary since, modern commercial FPGAs have different modes in their LUT usage or memory blocks. The architecture is described in XML, and the toolflow is supported by *AA-Pack* [10] for the packing of logic blocks and *VPR* 6.0 [9, 11] for the place and route part. Furthermore, the set of benchmarks that can be used with the toolflow are not restricted to well-known (but old) BLIF-format circuits, which are somewhat outdated and no longer completely representative of current and future applications. Thus, a logic synthesizer (*ODIN* [12]) has been connected to the flow, in order to input directly high-level VERILOG benchmarks of large applications.

This tool suite has been developed for FPGA research. Hence, the use of elementary logic blocks other than LUTs is not supported natively. While it is not in our interest to build a new toolflow from scratch, we will complement it to meet our requirements. This allows us to ensure good compatibility with the FPGA architecture and to guarantee its efficiency. The use of fixed interconnect topologies is not supported by VTR tools. We must thus add a specific tool to the flow: MPack. This tool handles the packing of logic cells into an MCluster set. As in the previous flow, the ABC tool will be used to perform the technology mapping [13]. While it is always possible to map the circuits onto a LUT target, specific libraries will be used for this mapping. This is of course required while considering the use of reconfigurable logic gates with a lower set of achievable functions. It also allows benchmarks to be mapped with specific logic gates. For example, a specific tool [14] can be connected to the flow in order to build a library of logic gates based on a very specific set of logic functions realized with ambipolar transistors.

The obtained flow is very versatile, since it is possible to evaluate several architectural scenarios. A set of addressable scenarios is depicted in Fig. 6.7. Figure 6.7a shows the traditional LUT-based FPGA scenario. Obviously, the original flow is still compatible with FPGAs, since it is based on the FPGA-dedicated VTR flow. Figure 6.7b corresponds to the architectural scheme, which has been proposed in this work. The flow addresses the matrix clustering and packing by the specific MPack tool. Then, since the architectural organization is

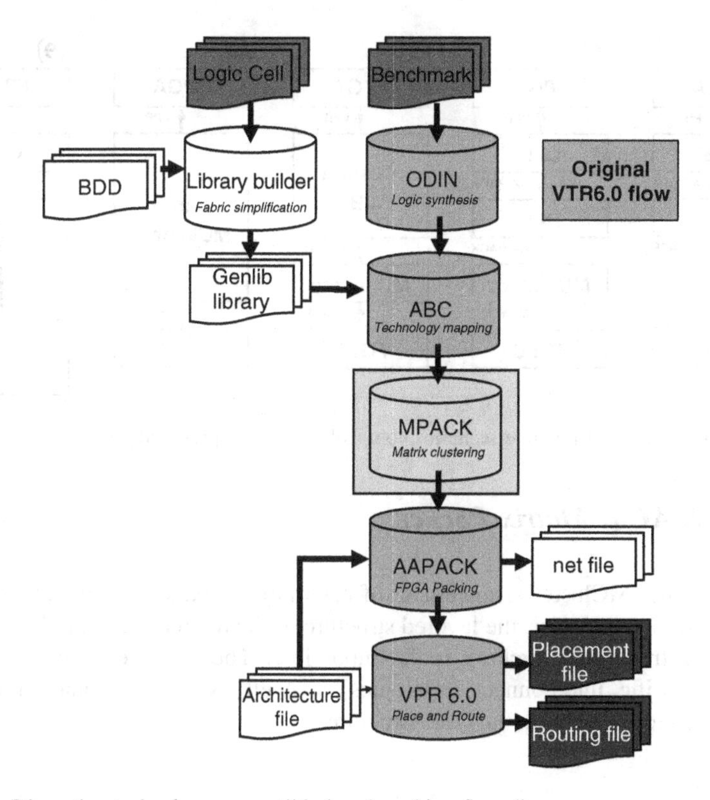

Fig. 6.6 Disruptive technology compatible benchmarking flow diagram

close to that of FPGAs, it is possible to pursue the packing and routing with the VTR toolflow. Figure 6.7c illustrates that, after matrix packing, it is possible to explore new kinds of organization for the high level logic layers of the structure. In particular, some opportunities for these levels will be discussed at the end of this chapter. Figure 6.7d shows that the toolflow is compatible with very regular approaches. These architectures have been proposed for computation architectures in environments with a high level of defects. Thus, it is possible to envisage the description of the NanoFabric organization [15] or the Cell Matrix approach [16], where large matrices of cells interconnected with their neighbourhood are used to create the logic plane. Finally, Fig. 6.7e depicts the ability of the flow to further consider reconfigurable architectures with specific custom logic instead of programmable elementary components. Hence, the flow is compatible with approaches, which propose a logic cell as the elementary combinational logic block [17, 18].

However, at the time of writing, it should be noted that *VPR* 6.0 is still under development regarding timing analysis. More particularly, the tool cannot model inter-CLB timings. This makes it impossible to carry out exact comparisons to other architectures regarding delay. This feature will be enabled later.

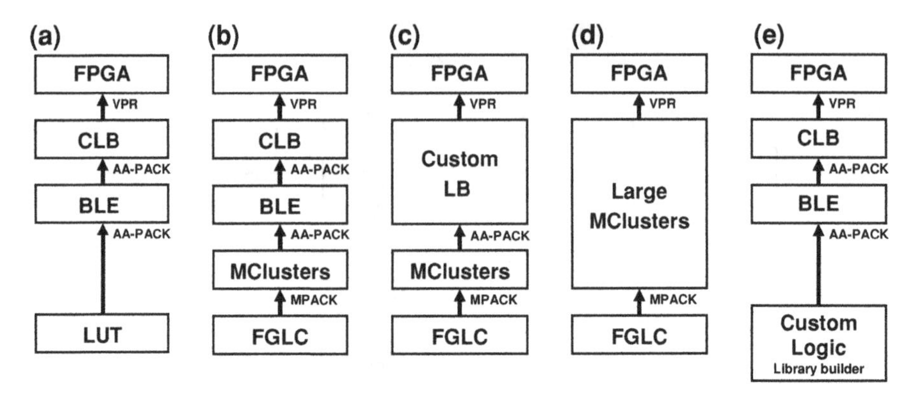

Fig. 6.7 Illustration of possible scenarios compatible with the proposed toolflow

6.3.2 MPACK: Matrix Packer

The use of the MCluster arrangement of cells introduced several novelties to the architecture. In particular, the layered structure of the interconnect topology makes the use of traditional packing tools impossible. The proposed tool is able to manage specific interconnect topologies, as well as the associated buffering generation required by the layered structure.

6.3.2.1 Concept

The Matrix Packer has been designed to pack netlists of logic cells into a set of MClusters. Figure 6.8 describes the internal organization of the packing tool MPack. The tool is composed of two principal algorithms: the mapper and the clusterer.

The mapping algorithm aims to fit a netlist of logic cells onto an MCluster architecture. This module is split into two parts: the architectural optimization and the brute force mapper.

The clustering algorithm aims to group the logic cells into arrangements of MClusters. These arrangements are subsequently mapped onto the architecture in order to validate the legitimacy of the structure.

To ensure the integration into the whole flow, netlists are read and exported using the BLIF file format [19]. The Architecture Generator block generates the MCluster template. Finally, in order to help with tool debugging, it is possible to export every manipulated graph graphically, using the DOT language [20].

The targeted MCluster shape is parameterized in terms of the size and topology scheme. An example of a targeted Banyan topology with a $d = w = 4$ is shown in Fig. 6.9a. In the tool internal representation, the connectivity between nodes is represented by the adjacency matrix of the graph. In such matrices, (i, j) refers to the

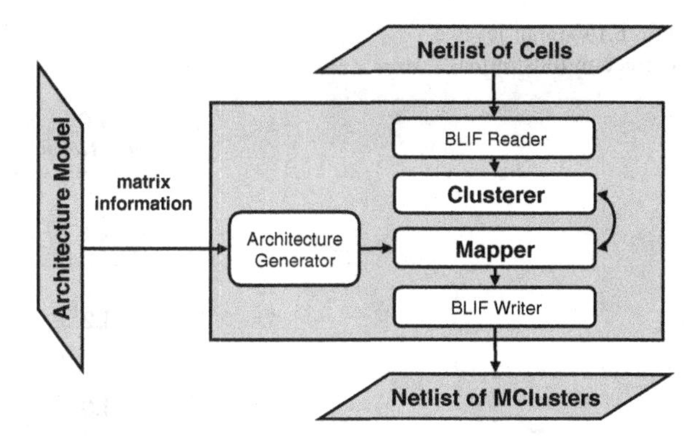

Fig. 6.8 MPack model flow

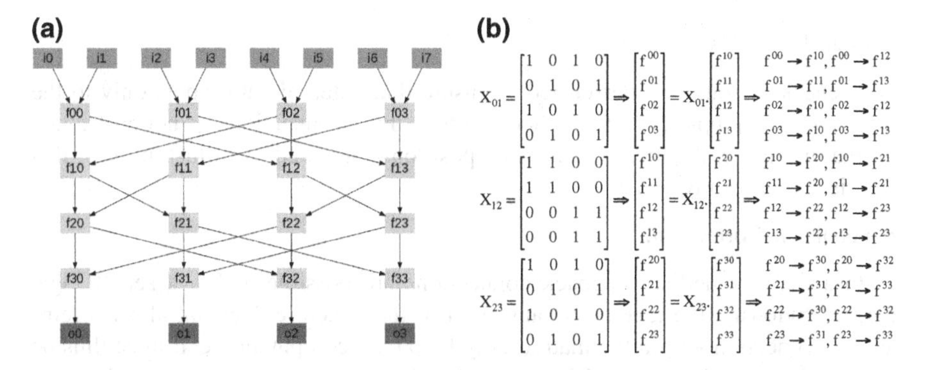

Fig. 6.9 **a** Banyan_4_4 topological arrangement (logic cells are labeled f*xx*, virtual input nodes i*xx* and virtual output nodes o*xx*) and **b** associated cross-connectivity matrices

intersection of row i and column j. A 1 at the position (i, j) means that the point i is connected to the point j. As an example, the individual cross-connectivity matrices X_{nm} (X_{01}, X_{12} and X_{23}) between logic cell stages of depth *n* to *m* are shown in Fig. 6.9b.

6.3.2.2 Mapping Algorithm: Architectural Optimization

In the mapping process, an architectural optimization is first performed in order to adapt the netlist to the physical architectural scheme. In fact, due to the layered structure, the logic can be seen as pipelined. The input netlist of cell graphs has to be processed by adding necessary synchronization elements to extend the input and output data paths, as well as by adapting connections to conform to the physical topology. For instance, in an ABC output netlist, any logic gate output can be connected to several gate inputs. However, MClusters have a fixed fan-in and fan-out. The netlist is made MCluster-compatible by adding buffering nodes to limit the fan-out. The netlist is thus processed as follows:

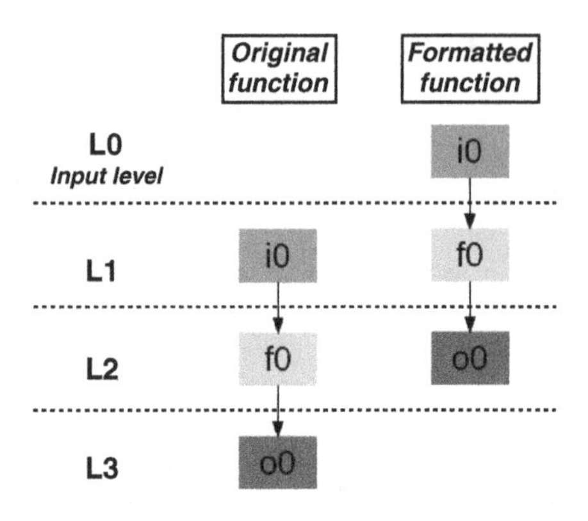

Fig. 6.10 Input level correction illustration

- Input Layer Correction:

  A check operation is performed to ensure that external data inputs only to the first layer. If an input node (i.e. a node with no input connections) is not positioned on the first layer, the hierarchical node position is updated. An illustration of this step is shown in Fig. 6.10.

- Feedback Correction:

  In a real-life netlist, feedback connections are possible within a set of logic gates. Obviously, the layered organization of the structure does not allow a retro active connection within the matrix. Any loop on a computation cell must thus be handled by the BLE external interconnect. Hence, it is necessary to modify any feedback connection into additional output and input terminals to ensure external connectivity. This case is illustrated in Fig. 6.11.

- Jump Correction:

  Due to the layered structure, a connection which "jumps" at least one logic layer is not allowed by the topology. In fact, a connection between two layers must pass through a logic cell. Consequently, the creation of a path (by the addition of a buffer cell) instead of a jump is necessary to handle such a situation, as shown in Fig. 6.12.

- Multiple Buffering/Inverting Path Correction:

  A logic cell which is connected to only one input signal is a buffer or an inverter. Hence, we can try to find multiple buffering/inverting paths in the logic function. Figure 6.13a depicts the situation of an incoming path. Two values f0 and f1 are propagated by two buffer paths (formed by f2 and f3 respectively). The signals are then used by the f4 cell to compute the o0 result. It is clear that buffering both inputs to propagate them through the structure is not optimal. It is of course preferable to perform the operation as soon as possible in the path and to buffer only the result, thus reducing the number of buffers by 2×. Figure 6.13b depicts the optimization of an outgoing path, which is analogous to the previously described case but is in terms of the output signals.

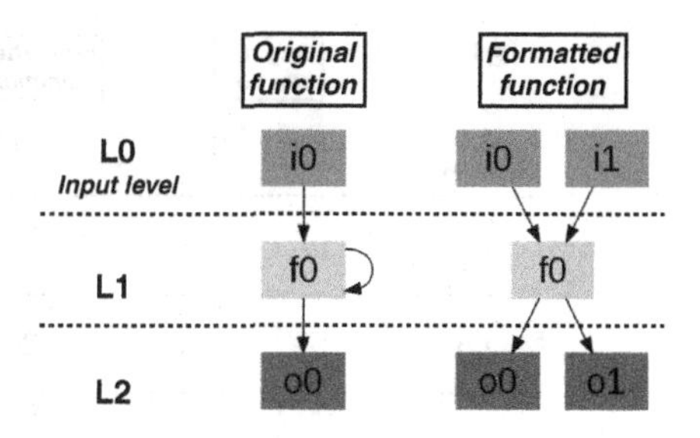

Fig. 6.11 Feedback correction illustration

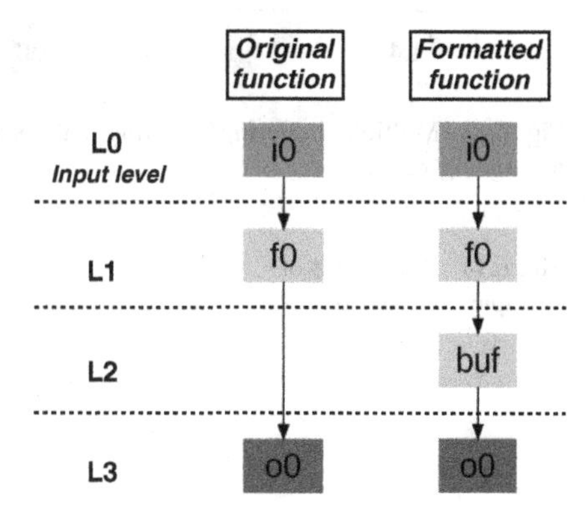

Fig. 6.12 Jump correction illustration

- Multiple Output Correction:

In the MCluster organization, the logic cells are connected to a limited number of other cells, i.e. fan-in/fan-out limitation. Hence, it is necessary to add buffers, in order to duplicate the signals in an architecturally compatible strategy. Figure 6.14 depicts an example case, where node f0 drives three different outputs. In this example, it is not possible to obtain more than two outputs per logic node. Hence, a buffer is added to reach the constraint.

- Output Layer Correction:

In this final optimization step, all the outputs of the logic functions are placed on the physical output layer. In Fig. 6.15, we consider a logic function where the outputs are on layers L2 and L3. To meet the constraints of the architecture (here a matrix of depth 4), buffers are added to place the outputs on the layer L5.

6.3.2.3 Mapping Algorithm: Brute-Force Mapper

After the logic functions have been processed to meet the specific architectural requirements, the logic function netlist is brute-force mapped to the architecture netlist. The adapted netlist is analyzed in depth, meaning that for each node in the structure, child branches are identified and recursively explored. This depth

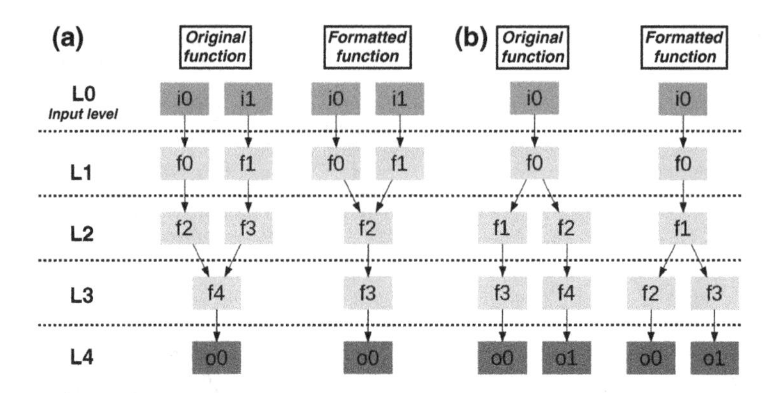

Fig. 6.13 Multiple buffering/inverting path simplification illustration.— **a** incoming path—**b** outgoing path

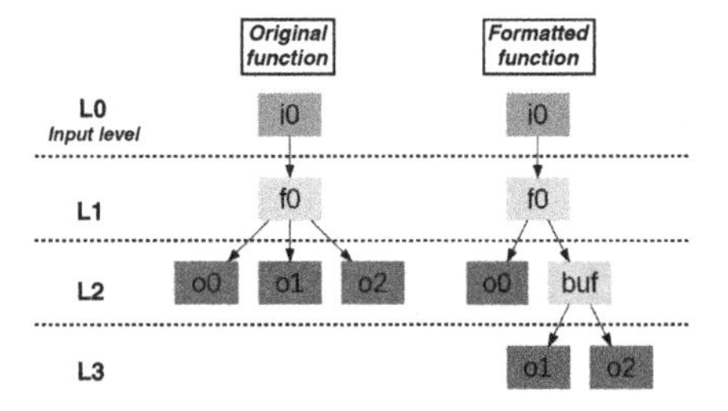

Fig. 6.14 Illustration of multiple output correction

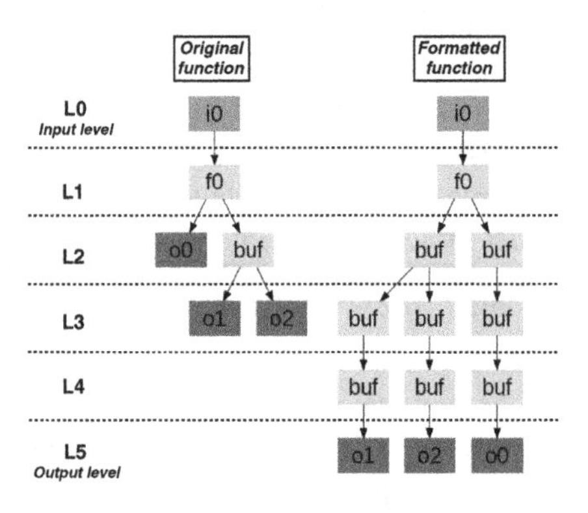

Fig. 6.15 Output level correction illustration

exploration is performed without any consideration of the edge orientation in the graph. This allows identification of the connection between the nodes and initiation of the mapping sequence. Logic nodes are then assigned to cells. This is done according to the physical interconnections. Each layer's connections are compared to the relevant inter-layer architectural connectivity—allowing (or not) the assignment of functions to cells. Branching (i.e. the exploration of the immediately preceding alternative) is used when the arbitrary choice leads to a dead-end; the

```
FUNCTION Mapping_procedure(current node)
  IF (all children/parents of current node already placed) THEN
    FOR (empty cell connected to children or parents) THEN
      Store the potential positions
    END FOR
  ELSE Store all the positions on the considered level
  END IF

  IF (No positions have been found) THEN
     MAPPING FAILURE
  END IF

  FOR (Each potential position)
     Place the current node at the position
     Ret=Mapping_procedure(Next node)
     IF (Ret is MAPPING FAILURE) THEN
       Unplace the current node from the position
     END IF
     IF (Ret is MAPPING SUCCESS AND No more nodes) THEN
       MAPPING SUCCESS
     END IF
  END FOR
```

Fig. 6.16 MPack mapping algorithm (pseudo-code)

process is repeated until all functions are assigned to cells. The recursive algorithm of this step is shown in Fig. 6.16.

As an example, we again consider a matrix which is four cells deep and four cells wide using a Banyan interconnect topology (Fig. 6.9a).

The example function to map is represented by a graph (shown in Fig. 6.17), generated by a random graph generator for test purposes. Its adjacency matrix is also shown in Fig. 6.17.

Following the previously described adaptation methodology, the original function graph is corrected in order to maximize the matching with the targeted MCluster. Hence, the position of inputs/outputs and inter-layer jumps will be corrected. The corrected graph is depicted in Fig. 6.18.

The graph exploration is then launched and we obtain the following sequence (obviously, *buf* represents synchronization nodes):

$$pi0 - p1 - pi1 - buf - p7 - p4 - p2 - pi2 - pi3 - p5 - p8 - buf - po1 - p9 \\ - buf - po2 - buf - po0 - pi4 - p3 - p6 - buf - buf - po3$$

Finally, the graph assignment is performed. In the example, the first point p1 is assigned to the cell f^{00}. According to the path defined in the previous step, a buffer is the next point to assign to a cell in the matrix. Since f^{00} is physically connected

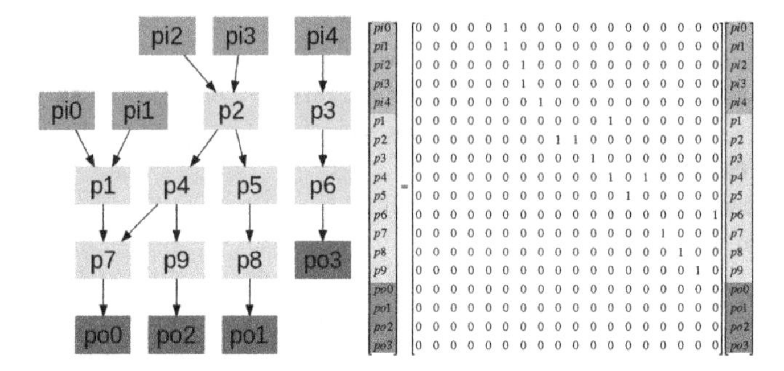

Fig. 6.17 Function graph to map onto the Banyan_4_4 interconnect topology and its associated adjacency matrix

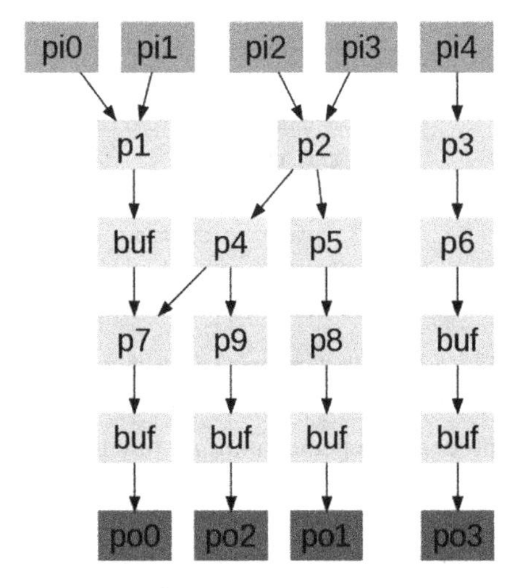

Fig. 6.18 Function graph after correction step

to f^{10} and f^{12}, the cell with lower y-index (here f^{10}) is arbitrarily chosen for the buffer assignment, and the other possibility is memorized. In our example, the final programmed matrix is shown in Fig. 6.19. In this figure, we can see the nodes of the logic function graph and the nodes added for synchronization purposes correctly placed on the cell matrix.

6.3.2.4 Clustering Algorithm

The clustering algorithm aims to group the logic cells into arrangements of MClusters. Such functionality is widely used in CAD tools.

There is large body of prior work on the packing problem for FPGAs. Most of it focuses on the optimization of area, delay, and/or power for the basic (LUT-based) soft logic complex blocks. These algorithms include T-VPack [21], T-RPack [22], IRAC [23], HDPack [24], and others [25–28].

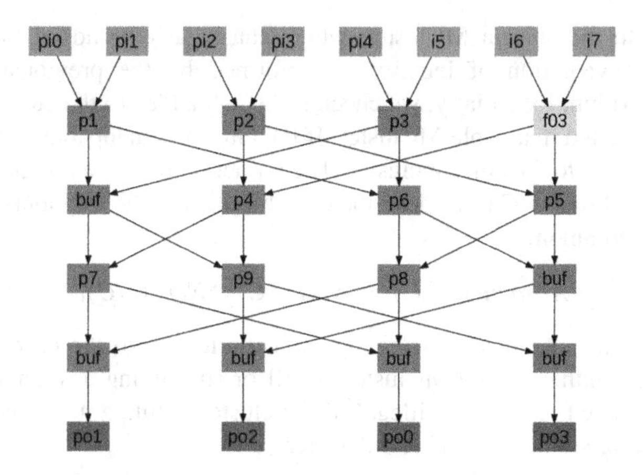

Fig. 6.19 MCluster after function packing

```
Remaining_Cells: set of unclustered logic cells
C: set of cells packed in the current MCluster
MClusters: set of MClusters

MClusters = NULL
WHILE (Remaining_Cells != NULL) // More Cells to cluster
  C = Seed (Remaining_Cells) // Most Used Inputs and legal Cell
  WHILE (|C| < (MCluster_w.MCluster_d)) // Cluster is not full
    Best_Cell = MaxAttractionLegalCell(C, Remaining_Cells)

    IF (Best_Cell == NULL)
      BREAK
    END IF

    Remaining_Cells = Remaining_Cells - Best_Cell
    C = C ∪ Best_Cell
  END WHILE
  MClusters = MClusters ∪ C
END WHILE
```

Fig. 6.20 MPack' clustering algorithm (pseudo-code)

In the MPack tool, we used algorithms derived from VPACK [29]. The tool constructs each MCluster sequentially, where the algorithm greedily packs the cells into MClusters. The pseudo-code for the algorithm is shown in Fig. 6.20.

It starts by choosing a seed cell for the current MCluster. As described by [29], the best way to choose the seed is to select the unclustered cell with the most used inputs. This approach prioritizes early placement of cells using the most cluster inputs, which are a scarce resource. Next, the tool selects the cell with the highest "attraction" to the current MCluster, which can legally be added to the current cluster. This evaluation of legality is performed by the previously described mapping algorithm. Essentially, we check if the cell added to the current MCluster leads to a valid and mappable MCluster. If the cluster is mappable, then the cell is definitively added to the current cluster. The attraction between a logic cell *LC* and the current MCluster *MC* is, as described in [29], the number of inputs and outputs they have in common:

$$Attraction\ (LC) = |Nets\ (LC) \cap Nets\ (MC)|$$

This procedure of greedily selecting the cells to add to the current MCluster continues until either (i) the MCluster is full or (ii) adding any unclustered cell would make the current cluster illegal. If the cluster is full, a new seed is selected and the packing starts for another MCluster.

6.4 Evaluation of Fixed Interconnect Topologies

This section aims to evaluate the impact of the choice of interconnect topology in the MCluster architecture. The topology is evaluated considering performance metrics in the context of the packing method. It is worth pointing out that the evaluations proposed in this section are based only on the packing method presented above. The results are then considered valid at the cluster level, and are used to validate the methodology. Large benchmarks, based on the complete tool flow, will be evaluated in the next section.

6.4.1 Methodology

Our analyses were carried out on an MCluster_4_4 using the previously mentioned intra-matrix interconnection topologies. We evaluated various metrics: the success rate of packing function graphs, the fault tolerance and the average interconnect length. We have carried out detailed analyses to compare the efficiency of the different intra-matrix interconnect topologies. We use a random graph generator to generate static sets of function graphs containing 6–16 points, in order to obtain fixed comparison criteria between topologies. No graphs contain isolated nodes, as here we focus on fixed interconnect layers, which are severely penalized by isolated nodes. Each set, which corresponds to a given number of points in the function graph, contains 1,000 samples. Using the previously described packing method, each function is programmed onto the MCluster_4_4 using the various intra-matrix interconnect topologies (ideal or faulty) and metrics are calculated. MCluster_4_4 was an arbitrary choice because there is a good tradeoff between complexity and simulation time. Figure 6.21 summarizes the evaluation methodology and the associated parameters.

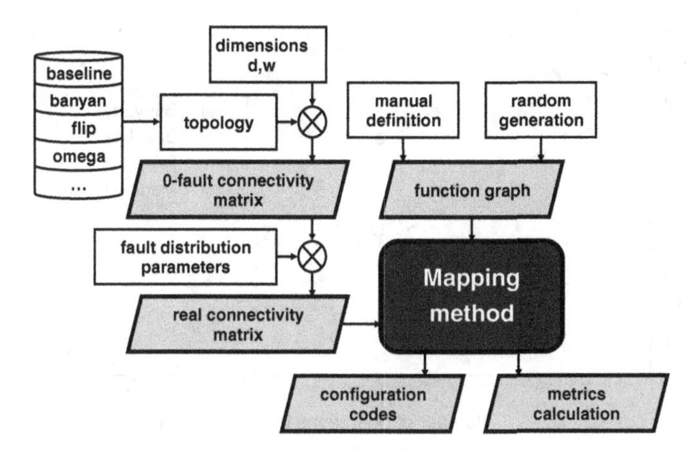

Fig. 6.21 Performance evaluation method for fixed interconnection topologies

6.4.2 Packing Success Rate

Applying static sets to ideal interconnect topologies, we can test the ability of the matrix-topology ensemble to have complex functions packed onto it. Considering the percentage of function graphs successfully packed onto the matrix with respect to the number of samples in the set, we obtain the *success rate*. Figure 6.22 shows the comparison of success rates for Banyan, Modified Omega, Flip and Baseline topologies. For Banyan, Flip and Baseline interconnect topologies, the success rate is about 80% when the function graphs have six points. At 12 points, the success rate is about 25%. The difference between these three topologies is thus relatively small. However for the Modified Omega interconnect topology, the success rate is about 90% for six-point function graphs and about 40% for 12-point graphs. This clearly shows that the Modified Omega interconnect topology is more suitable for this type of matrix.

This is because this topology has lower redundancy than the other topologies and spreads calculations over cells occupying less width, which seems to correspond better to typical function graphs. In fact in the matrix, there are pairs of cells which have the same inputs. For two cells which have the same inputs, the sum of the number of functions they can achieve is 14. For two cells which do not have the same inputs, the sum of the number of functions they can implement is $14 + 14 = 28$. In the Banyan topology for example, there are six pairs of cells which have the same inputs, while in the Modified Omega topology, every inputs are different. This is the main reason why the Modified Omega topology has the potential to realize more functions than other topologies.

It is worth noticing that a traditional LUT approach would lead to a 100% mapping success rate, insomuch as all possible combinational functions could be achieved by a LUT with a sufficient number of inputs. Then, the use of a MIN reduces the number of packed logic functions. Such a problem is managed at a

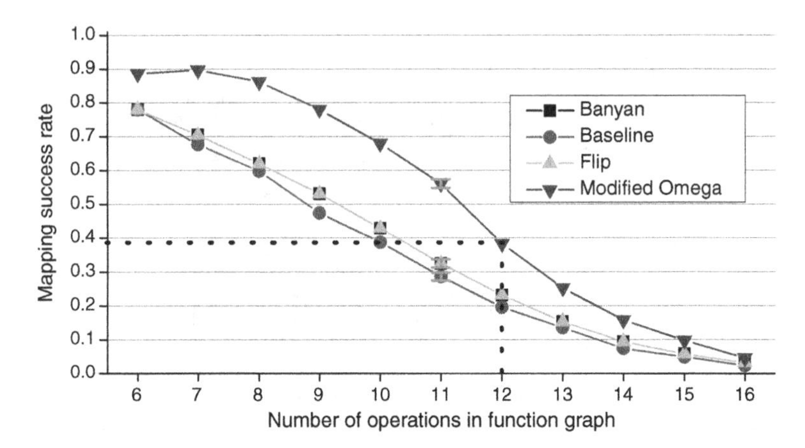

Fig. 6.22 Programmability success rates for Banyan, Modified Omega, Flip and Baseline interconnect topologies within 4-deep 4-wide matrices

higher hierarchical level. For example, if a function graph cannot be packed onto a single matrix, we can split it and map the sub graphs onto different matrices.

6.4.3 Fault Tolerance

6.4.3.1 Physical Origin of Defects at Nanoscale

A significant concern in the use of nano-scale devices for computing architectures is the reliability of these individual devices and that of the resulting system. For instance, the chemical processes for building devices in a bottom-up approach will have significantly lower yields than those obtained via current fabrication practices, resulting in aggregates with high defect rates. For example, when considering transistors based on carbon nanotubes, defects introduced by the nanotube synthesis process could impact the behavior of the CNFET [30]. It is also necessary to consider that traditional processes have to shrink to build such devices. Due to variability, the overall system performance is thus likely to be dominated by unreliability as concerns individual device characteristics. Thus, it will be utopian to consider that interconnection layers will be defect-free. Most probably, at least one link could be destroyed or stuck in a Boolean state [31].

6.4.3.2 Analysis Protocol and Results

Typically, unreliable systems use defect-avoidance techniques [32] to increase their reliability. Such techniques are based on defect detection and are packing-aware.

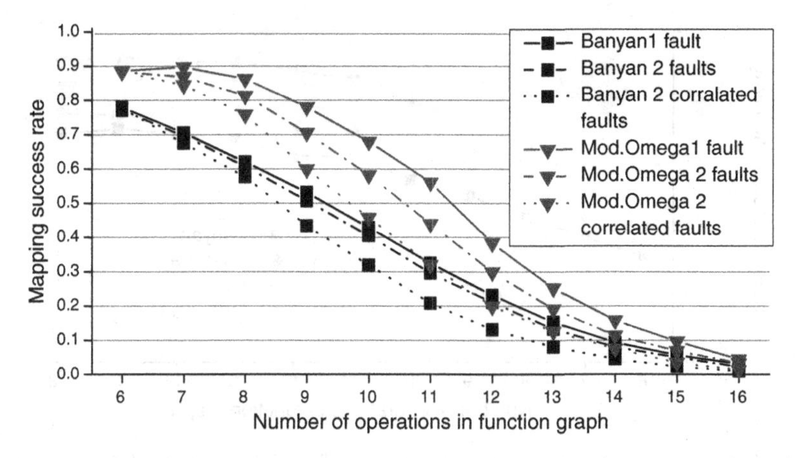

Fig. 6.23 Comparison of programmability success rates for Banyan and Modified Omega interconnect topologies within 4-deep 4-wide matrices in the case of faulty links and faulty cells

To study fault tolerances in matrices, we must introduce defects in the interconnection layers, which will force the packing method to work around them. This will simulate the behavior we could achieve with a defect avoidance technique. With these considerations, physical interconnection defects were simulated by randomly deleting links in layers. Cell non-functionality is simulated by deleting all related input/output links. By so-doing, we assume that the faults are only static and due to fabrication processes issues, i.e. they do not change dynamically during runtime. It is also worth noticing that, by using defect-avoidance techniques, we assume that the faults are testable.

Using the Banyan topology as a reference, Fig. 6.23 shows the comparison of success rates for Modified Omega and Banyan topologies in the cases of one or two faulty links on the first layer and one faulty cell, which are the most representative cases. For all topologies, when the function graphs have six points or 16 points, faults have no influence on the success rate, because the interconnect topologies are not a decisive factor at these limits. For 12 points, the success rate of the Modified Omega topology falls faster than for the Banyan topology. The Banyan topology can thus be considered to be more robust than the Modified Omega topology in terms of sensitivity to faults, but in all situations, the absolute success rate for the Modified Omega topology remains higher than that of the Banyan topology.

6.4.4 Average Interconnect Length

Arbitrary coefficients are assigned to interconnect based on how long they are. Given an interconnect between cells at coordinates (x, y_1) and (x + 1, y_2) respectively, the coefficient value is given by $|y_2 - y_1| + 1$ (assuming a Manhattan-based interconnect geometry).

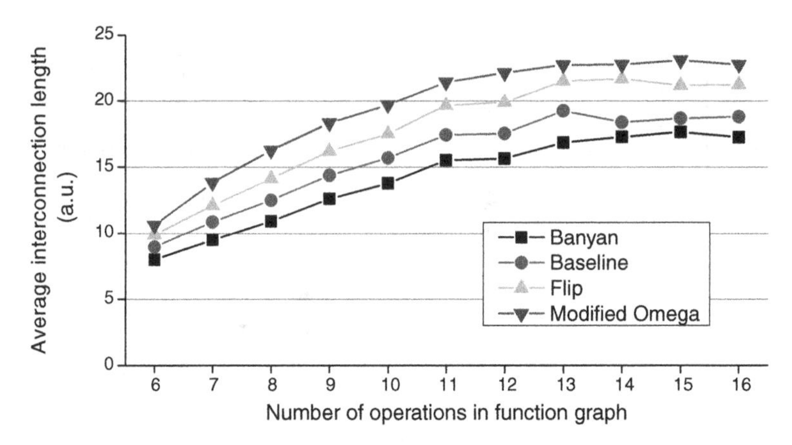

Fig. 6.24 Comparison of average interconnect length for Banyan, Modified Omega, Flip and Baseline interconnect topologies within 4-deep 4-wide matrices

Directly after the packing process, it is therefore possible to sum the coefficients of all used interconnects. This method gives information about the interconnect length used in the system. Considering layout and design rules, it is possible to extract parasitic capacitances, and therefore to deduce the maximum working speed of one pipeline stage.

Figure 6.24 shows the average interconnect length comparison. The Banyan topology displays the lowest interconnection length due to its straight-ahead paths. However, the Modified Omega topology, with its long shuffle links, is the worst case with 25% greater overall interconnects than Banyan. This shows that the average system speed is likely to be higher with the Banyan topology than with the others.

6.4.5 Discussion

The results, obtained above, are summarized in Table 6.1. The metric "Performance" is calculated as the inverse of the average interconnection length and all the values are normalized considering the use of the Banyan topology for a 12-node graph as reference, such that

$$Metric_{Normalized}(X) = \frac{Metric_{at\,12\,nodes}(X)}{Metric_{at\,12\,nodes}(Banyan)}.$$

The Modified Omega topology gives the best results in terms of success rate and fault tolerance, whereas it appears to be the worst for performance. However, considerations about performance will require deeper analyses considering the layout, and real values of extracted parasitic capacitances. The three other topologies analyzed are quite similar in results. As mentioned above, these results

Table 6.1 Analysis summary

	Packing efficiency		Performance (1/int. length)
	No faults	With faults	
Banyan	1	1	1
Baseline	0.84	0.86	0.89
Flip	1	1	0.78
Modified Omega	1.65	1.42	0.71

might be explained by the redundancy of the interconnect topologies. In fact, highly redundant topologies will give good results in terms of fault tolerance, while their packing success rate is less convincing.

6.5 Global Architecture Evaluations

6.5.1 Methodology

In this section, we evaluate the performance of the proposed architectural scheme. We will use the previously described toolflow and map the MCNC and ISCAS'89 benchmarks [33] to the target architecture. Different scenarios will be used to evaluate the impact of the granularity and the latch depopulation, and to compare the global performance with the CMOS counterpart. The CMOS reference architecture is formed by CLBs organized with ten BLEs of four-input LUTs. Twenty two inputs will then come from the external routing structures into the block. The physical parameters of the different structures are extracted using 65 nm Back-End data. The used metrics are the area and the critical routing delay. These metrics are computed during the Place and Route iteration of the flow. The area corresponds to the sum of the logic area, i.e. the area of used CLBs, and the routing area, i.e. the area of the used routing resources. These values are expressed using the λ parameter, which is defined as half of the minimum gate length. All areas are normalized with respect to the largest circuit implementation. The critical routing delay corresponds to the most constrained delay through the routing structures, i.e. the longest path external to the CLBs.

6.5.2 Evaluation Results

6.5.2.1 Impact of the Granularity

Figure 6.25 depicts the area estimation of an FPGA, following the proposed architectural scheme. While the MClusters are totally populated in terms of latches, we study the impact of their granularity by varying the size of the clusters. The best

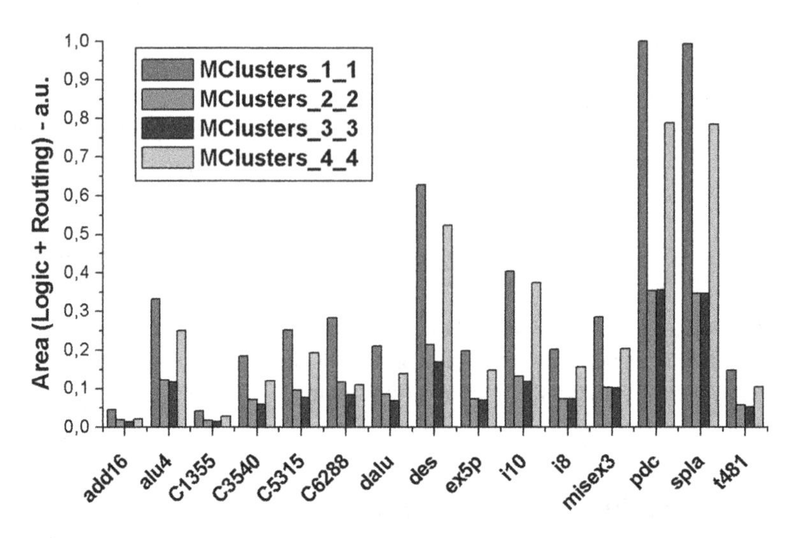

Fig. 6.25 Area estimation for MCluster-based FPGAs with various granularities

results are obtained with MClusters_3_3, where an improvement of 12% on average, compared to a 2 by 2 granularity, can be observed. It is worth noticing that for a smaller, as well as for a higher granularity, the used area increases drastically. For large clusters, this can be explained by the large size of MClusters with respect to their poor mapping ability. In such cases, the incomplete interconnectivity becomes a major hurdle, because of the large number of wasted cells used only for buffering purposes. On the other hand, when a 1 by 1 granularity is used, the amount of external circuitry required for the routing heavily dominates the area, as compared to the computation cell. This leads to a large area overhead, and confirms the initial hypothesis of this work, where the direct transposition of an FPGA architecture to ultra-fine grain would lead to a large interconnection overhead.

6.5.2.2 Impact of Latch Depopulation

Figure 6.26 depicts the area estimation of the MCluster-FPGA with a 3 by 3 granularity and latch depopulation.

The depopulation ratio corresponds to the number of latches that have been removed in the structure, with respect to the maximum number of possible latches. We can observe that the use of latch depopulation is not a good strategy for area performance, because of a nearly 2× increase in used area. While the idea of removing some latches is interesting to handle the size of the cluster (the size of a latch is much greater than that of an elementary cell), it appears detrimental to the routing efficiency. In fact, the position of the latch is not taken into account by the MCluster packing tool. This leads to a less efficient placement iteration, and consequently to the inefficiency of the depopulation.

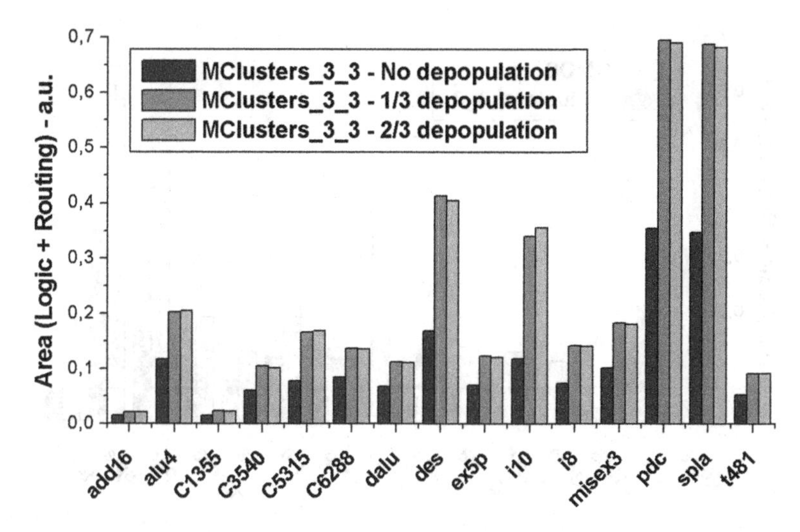

Fig. 6.26 Area estimation for MCluster_3_3-based FPGAs with different latch depopulation scenarios

6.5.2.3 Performance Comparison with CMOS

Figure 6.27 depicts the area estimation for a 3 by 3 granularity MCluster-based FPGA and compares it to its CMOS counterpart. The benchmarks show an area reduction ranging from 38 to 62%, with an average of 46%. This is clearly due to the low area impact of ultra-fine grain logic cells, compared to the rather larger area required by a CMOS LUT. In fact, a 3 by 3 cluster costs an area of $63{,}433\lambda^2$ (with λ the half of the minimum gate length) while a four-input LUT costs $54{,}292\lambda^2$. It is worth noticing that, while their size remains in the same order of magnitude, the functionality of an MCluster is much higher than that of an equivalent LUT. With the same size, MClusters can input 50% more data and can output 3$\times$ more results. When correlated with the efficiency of the packing tool for matrix clustering, this demonstrates a clear advantage of the complete proposal as compared to the CMOS approach.

Figure 6.28 depicts the distribution of the critical routing delay through all the benchmarks using three different scenarios. While the average delay is not changed significantly (delay reduction up to 10%), it is worth noticing that the use of MCluster structures allows a reduction in the standard deviation of the delay distribution. In fact, we can observe that, while the CMOS distribution is quite large, the use of a 2 by 2 MCluster implies a lowering of the extremes, and tends to globally improve the performance of the mapped circuits. This behavior can be explained through the global impact of ultra-fine granularity on the benchmarking toolflow. In fact, ultra-fine granularity induces a predominance of local inter-CLB interconnect instead of long wire connections, thus leading to the reduction of long critical paths.

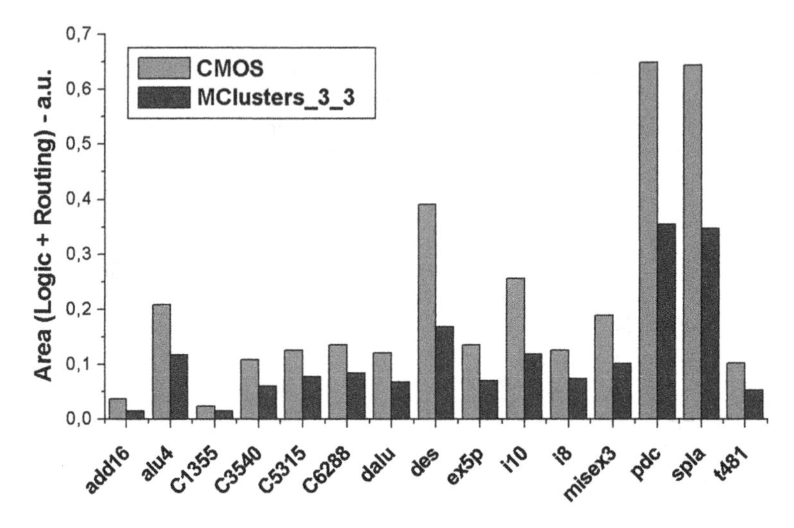

Fig. 6.27 Area estimation for 3 by 3 MCluster-based and CMOS-based FPGAs

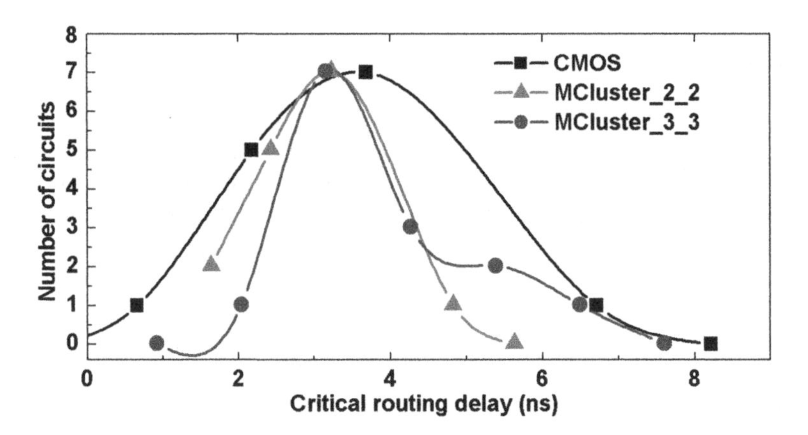

Fig. 6.28 Critical routing delay repartition for 2 by 2, 3 by three MCluster-based and CMOS-based FPGAs

6.5.3 Discussion

In this chapter, we proposed a layered structure, derived from conventional FPGAs. To build a fine-grain combinational block from a population of ultra-fine grain logic elements, it was necessary to add a new design layer.

In the envisaged architecture, ultra-fine grain logic cells are arranged into matrices, called MClusters. To prevent inherent interconnect overheads, the cells are interconnected through fixed and incomplete topologies. Several topologies could be envisaged. The MClusters are further organized into BLE/CLBs,

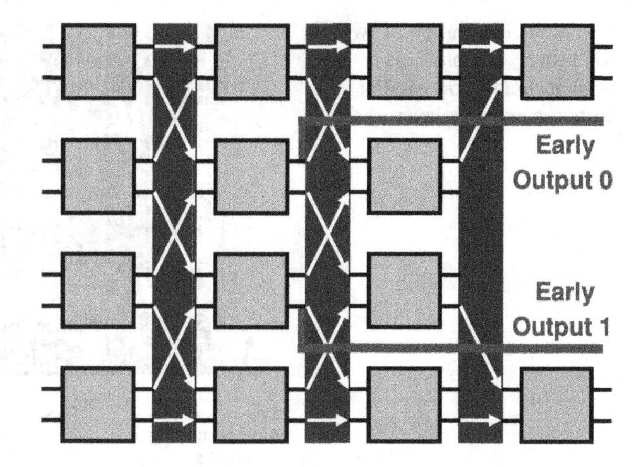

Fig. 6.29 Modified-Omega based MCluster with early internal outputs and cell depopulation

with complete interconnect and the CLBs are arranged in islands to form the FPGA. The Modified Omega interconnect topology gives the best results in an MCluster_4_4 context.

It is worth noticing that several other topologies can be investigated. In other work, a cross-cap pattern has been proposed for the matrix arrangements [34, 35] to overcome logic depth and data directivity concerns. Another solution to constant logic depth, where the layered matrix structure requires a large number of buffers to correctly output results, is to modify the structure to output some of the internal nodes of the matrix and delete some cells often used for only buffering, as shown in Fig. 6.29.

From an integration perspective, it is worth highlighting that a matrix organization is particularly well-suited to the structural regularity required by CNFET fabrication processes. Matrix patterns fundamentally exhibit structural regularity, which is easily transposable to highly regular fabrication processes. For example, in the case of in-field reconfigurable cell, even if there are many fundamental CNFET integration issues which are still unsolved, the architecture is compatible with recent proposals and fabrication results [36, 37], since it is possible for long bundles of nanotubes to span several cells in the same column (Fig. 6.30).

From a higher-level perspective, the MCluster organization has demonstrated good improvements in area and delay with respect to CMOS when using a 3 by 3 granularity.

While the proposed solution appears particularly interesting for future FPGAs, it will also be of high interest to study other architectures, which go beyond conventional CLB organization. In fact, the approach presented in this chapter is quite close to the traditional FPGA scheme. In Figs. 6.31 and 6.32, we present a more speculative proposition for the *Configurable Nano Logic Block* (CNLB). Figure 6.31 shows an MCluster_4_4 connected to four sequential elements, dedicated to latch the outputs of the cluster. Peripheral circuitries are used to allow serial loading of configuration data, thus reducing the number of interconnect metals lines that are used for configuration. Figure 6.32 shows the organization of

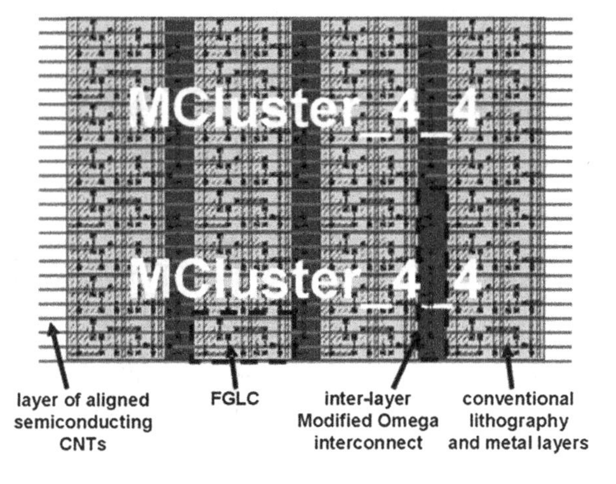

Fig. 6.30 Illustration of two MClusters_4_4 physical implementation on parallel carbon nanotubes layer

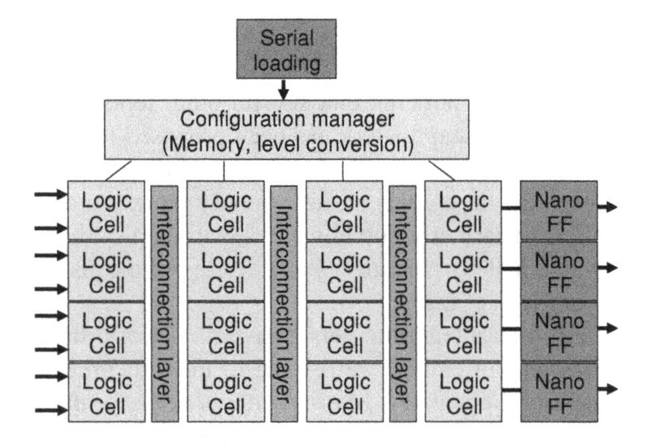

Fig. 6.31 MCluster organization for speculative CNLB

the CNLB. We propose the use of a logic pipeline. In this arrangement, the logic structure presented in Fig. 6.31 is duplicated many times and cascaded into many pipelined paths. A configurable interconnection is possible between the pipelines. This organization can be considered as another layered arrangement, similar to that of the MCluster. Furthermore, management circuitry will be added to this circuit, to ensure service tasks, such as scheduling or synchronization between tasks and paths through the pipeline, as well as ensuring the correct behaviour of the whole. For example, a defect mapping operation could be in charge of a dedicated state-machine-based test circuitry to return the information to upper-level blocks in charge of defect avoidance.

Another final possibility is also to pursue the straightforward MCluster organization. For this case, Fig. 6.33 shows a double-stage MCluster organization with four MClusters_4_4 interconnected using a Modified Omega topology (intra-matrix interconnect). The four internal MClusters are then packed into another MCluster organization of two blocks in depth and two in width. These blocks are

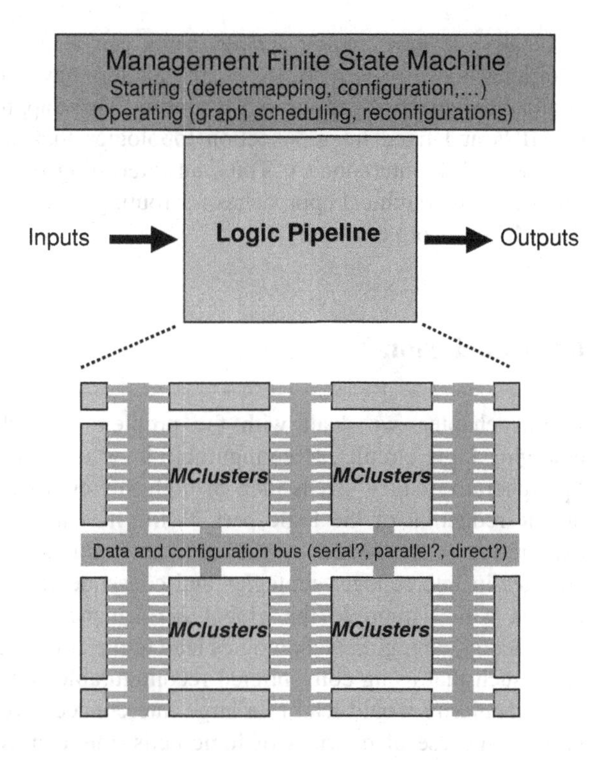

Fig. 6.32 Speculative pipelined organization of CNLB

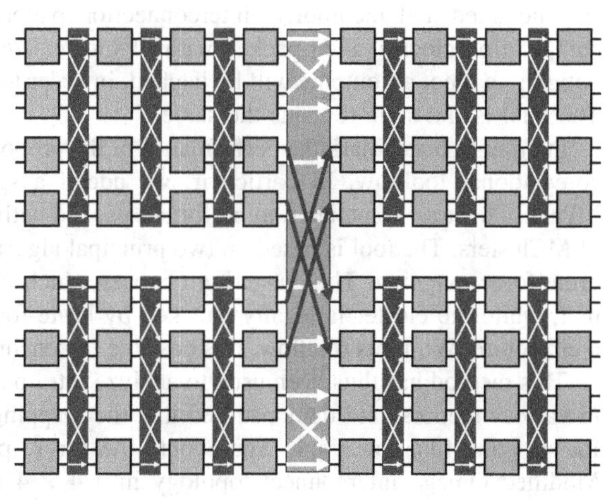

Fig. 6.33 Double-stage MCluster organization with two-level hierarchic interconnection strategy

then interconnected by a Modified Banyan topology (white inter-matrix interconnect), which aims to interconnect the various blocks from side to side (black inter-matrix interconnect), while maximizing direct interconnections. This final proposal helps to show that the design exploration space, reachable by the proposed benchmarking flow for ultra-fine grain, is both broad and promising.

Finally, several routing arrangements can also be envisaged instead of the traditional mesh pattern. In [5, 6], the authors have proposed a convenient hierarchical organization for intra-CLB routing. These approaches are based on the use of MINs and fixed interconnection topologies that could efficiently replace the complete CLB interconnect. Thus, an interesting opportunity could be found in mixing the hierarchical approaches for routing (i.e. hierarchical routing) and logic (i.e. MCluster organization).

6.6 Conclusion

In this chapter, we dealt with the problem of architectural organization for emerging logic circuits. Reconfigurable logic uses the largest part of its area for "peripheral circuitry". It is then of high interest to explore the implication of a drastic reduction of the logic part. While it is not obviously of first order, this exploration is motivated by the fact that reducing the logic part may lead to changes in interconnect strategies and consequently to new architectural organizations which improve the global performance of the reconfigurable device. Specifically, emerging technologies lead to the emergence of highly compact logic cells, such that using conventional reconfigurable computing schemes with ultra-fine granularity would lead to a large interconnect overhead. To prevent this, we propose the use of matrices of logic cells, interconnected by layers and using a specific fixed and incomplete interconnection pattern. This new combinational computation block is then packed with FFs to realize a BLE scheme. It is worth noticing that not all outputs will be latched, in order to reduce the size of the BLE. The final island-style remains the same.

In order to benchmark the performance of the proposed circuit, we enhanced the conventional toolflow. In particular, we added a specific packing tool, called MPack. This tool is intended to perform a packed netlist of logic cells into netlists of MClusters. The tool is based on two principal algorithms: greedy clustering and brute force mapping. The clustering chooses which blocks must be packed together, while the cluster is legally checked by brute-force mapping onto the target architecture. With this toolflow, it is possible to benchmark the proposed structure.

This method has thus been used to analyze intra-matrix topologies with respect to various metrics, and demonstrated that the mapping success rate is about 90% for six-point function graphs and about 40% for 12-point graphs when using the Modified Omega interconnect topology in a 4×4 matrix. Moreover, we have shown that the Modified Omega interconnect topology is still more robust than other topologies in the presence of defects. In contrast, the Modified Omega topology is not a good choice for speed, displaying 25% greater overall interconnect length within the cell than the Banyan topology.

We subsequently evaluated the potential of the architecture in terms of MCluster granularity and latch depopulation, and compared the structure to a CMOS FPGA. We showed that the best granularity for used area is a 3 by 3 cluster, while the latching remains entirely populated. We finally showed an improvement with respect to CMOS with an average area saving of 46%, and we discovered that the routing delay is less distributed and tends to be more controllable than in the traditional approach.

To summarize, ultra-fine grain architectures appear to be highly relevant for future computation systems. Furthermore, many logic cell arrangements and interconnect topologies are possible at several levels. With the help of complete architectural exploration tools, the field could reveal interesting opportunities for future systems.

References

1. M. Lin, A. El Gamal, Y.-C. Lu, S. Wong, Performance benefits of monolithically stacked 3-D FPGA. IEEE Trans. Comput. Aided Des. Integr. Circuits Syst. **26**(2), 216–229 (2007)
2. E. Ahmed, J. Rose, The effect of LUT and cluster size on deep-submicron FPGA performance and density. IEEE Trans. Very Large Scale Integr. (VLSI) Syst. **12**(3), 288–298 (2004)
3. T. Ye, L. Benini, G. De Micheli, Packetization and routing analysis of on-chip multiprocessor networks. J. Syst. Architect. **50**(2–3), 81–104 (2004)
4. D.L. Lewis, S. Yalamanchili, H.-H. S. Lee, High performance non-blocking switch design in 3D die-stacking technology. In:*Proceedings of the 2009 IEEE Computer Society Annual Symposium on VLSI*, (2009), pp. 25–30
5. H. Mrabet, Z. Marrakchi, P. Souillot, H. Mehrez, Performances improvement of FPGA using novel multilevel hierarchical interconnection structure. In:*International Conference on Computer-Aided Design (ICCAD'2006)*, November 2006, pp. 675–679
6. Z. Marrakchi, H. Mrabet, C. Masson, H. Mehrez, Performances comparison between multilevel hierarchical and mesh FPGA interconnects. Int. J. Electron. **95**(3), 275–289 (2008)
7. C.-L. Wu , T.-Y. Feng, On a class of multistage interconnection networks. IEEE Trans. Comp. **29**(8), 694–702 (1980)
8. G.B. Adams, D.P. Agrawal, H.J. Siegel, A survey and comparison of fault-tolerant multistage interconnection networks. Computer **20**(6), 14–27 (1987)
9. Verilog-To-Routing (VTR) Project, http://www.eecg.utoronto.ca/vtr/
10. J. Luu, J. Anderson, J. Rose, Architecture description and packing for logic blocks with hierarchy, modes and complex interconnect. In:*ACM/SIGDA 19th International Symposium on Field Programmable Gate Arrays*, Monterey, 2011
11. J. Luu, I. Kuon, P. Jamieson, T. Campbell, A. Ye, M. Fang et al., VPRVPR 5.0: FPGA CAD and architecture exploration tools with single-driver routing, heterogeneity and process scaling. In:*ACM Symposium on FPGAs, FPGA '09*, February 2009, pp. 133–142
12. ODIN-II: Verilog synthesis tool, https://code.google.com/p/odin-ii/
13. ABC: Berkeley logic synthesis tool, http://www.eecs.berkeley.edu/~alanmi/abc/
14. M. De Marchi, M. H. Ben Jamaa, G. De Micheli, Regular fabric design with ambipolar CNTFETs for FPGA and structured ASIC applications. In:*International Symposium on Nanoscale Architectures (NANOARCH)*, June 2010
15. S.C. Goldstein, M. Budiu, NanoFabrics: spatial computing using molecular electronics. *28th Annual International Symposium on Computer Architecture*, (2001), pp.178–189
16. N.J. Macias, L.J.K. Durbeck, Adaptive methods for growing electronic circuits on an imperfect synthetic matrix. Biosystems **73**(3), 173–204 (2004)

17. M.H. Ben Jamaa, P.-E. Gaillardon, S. Frégonèse, M. De Marchi, G. De Micheli, T. Zimmer, I. O'Connor, F. Clermidy, FPGA design with double-gate carbon nanotube transistors. Electro Chem. Soc. Trans. **34**(1), 495–501 (2011)
18. J. Birkner, A. Chan, H.T. Chua, A. Chao, K. Gordon, B. Kleinman et al., A very-high-speed field-programmable gate array using metal-to-metal antifuse programmable elements. Microelectron. J. **23**(2), 561–568 (1992)
19. E.M. Sentovich, K.J. Singh, L. Lavagno, C. Moon, R. Murgai, A. Saldanha et al., SIS: a system for sequential circuit synthesis. *Technical Report UCB/ERL M92/41*, Electronics Research Lab, University of California, Berkeley, CA 94720, May 1992
20. DOT language—Graphviz tool,: http://www.graphviz.org/Documentation.php
21. A.S. Marquardt, V. Betz, J. Rose, Using cluster-based logic blocks and timing-driven packing to improve FPGA speed and density. In:*ACM/SIGDA 7th International Symposium on Field Programmable Gate Arrays*, (1999),pp.37–46
22. E. Bozorgzadeh, S. Memik, X. Yang, M. Sarrafzadeh, Routability-driven packing: metrics and algorithms for cluster-based FPGAs. J. Circuits Syst. Comp. **13**, 77–100 (2004)
23. A. Singh, G. Parthasarathy, M. Marek-Sadowksa, Efficient circult clustering for area and power reduction in FPGAs. ACM Trans. Des. Autom. Electron. Syst. **7**(4), 643–663 (2002)
24. D. Chen, K. Vorwerk, A. Kennings, Improving timing-driven FPGA packing with physical information.In: *International Conference on Field Programmable Logic and Applications*, (2007), pp. 117–123
25. G. Lemieux , D. Lewis, *Design of Interconnection Networks for Programmable Logic* (Kluwer Academic Publishers, New York, 2004)
26. J. Lin, D. Chen , J. Cong, Optimal simultaneous mapping and clustering for FPGA delay optimization. *ACM/IEEE Design Automation Conference*, (2006), pp. 472–477
27. A. Ling, J. Zhu, S. Brown, Scalable synthesis and clustering techniques using decision diagrams. IEEE Trans. CAD **27**(3), 423 (2008)
28. K. Wang, M. Yang, L. Wang, X. Zhou, J. Tong, A novel packing algorithm for sparse crossbar FPGA architectures. In:*International Conference on Solid-State and Integrated-Circuit Technology*, (2008), pp. 2345–2348
29. V. Betz, J. Rose, A. Marquart, *Architecture and CAD for Deep-Submicron FPGAs* (Kluwer Academic Publishers, New York, 1999)
30. P. Lambin, A.A. Lucas, J.C. Charlier, Electronic properties of carbon nanotubes containing defects. J. Phys. Chem. Solids **58**(11), 1833–1837 (1997)
31. E.K. Vida-Torku, W. Reohr, J.A. Monzel, P. Nigh, Bipolar, CMOS and BiCMOS circuit technologies examined for testability. In:*34th Midwest Symposium on Circuits and Systems*, 14–17 May 1991, pp. 1015–1020
32. J.R. Heath, P.J. Kuekes, G.S. Snider, R.S. Williams, A defect-tolerant computer architecture: opportunities for nanotechnology. Sciencel **280**(5370), 1716–1721 (1998)
33. BLIF circuit benchmarks,: http://cadlab.cs.edu/~kirill/
34. N. Yakymets, I. O'Connor, Matrice interconnectee de cellules logiques reconfigurables avec une topologie d'interconnexion croisee. Patent FR0958957, (2010)
35. N. Yakymets, K. Jabeur, I. O'Connor, S. Le Beux, *Interconnect Topology for Cell Matrices Based on Low-Power Nanoscale Devices, Faible Tension Faible Consommation* Marrakech, Morocco, May 30–June 1 (2011)
36. A. Lin, N. Patil, H. Wai, S. Mitra, H.-S.P. Wong, A metallic-CNT-tolerant carbon nanotube technology using asymmetrically-correlated CNTs (ACCNT). In:*Symposium on VLSI Technology*, 16–18 June 2009, pp.182–183
37. N. Patil, J. Deng, A. Lin, H.S.-P. Wong, S. Mitra, Design methods for misaligned and mis-positioned carbon-nanotube-immune circuits. IEEE Trans. Computer Aided Design **27**(10), 1725–1726 (2008)

Chapter 7
Conclusions and Contributions

Abstract In this chapter, we aim to conclude the book. A complete description of the research contributions and their impact on the nanoarchitecture community is given.

The global objective of this book was to provide a framework, suited for the comparison of different emerging technologies. In this concluding chapter, we summarize the main results and overall contributions of the previous chapters.

7.1 Global Conclusions

In this book, we intended to evaluate the potential of emerging technologies for future computing application. This topic is of course so broad that significant effort will be invested in the costly development of emerging technologies even if the system perspective cannot be foreseen. In order to reduce the lead-time and the costs, we explored a fast evaluation methodology. This methodology is based on a generic architectural template (Chap. 1). This was chosen to be a generalized FPGA architecture, since its high flexibility makes the circuit well-suited to several applications. After building a global overview of current and emerging reconfigurable architectures, we formalized the architectural template that will be the baseline of our study (Chap. 2).

The standard reconfigurable architecture drawbacks are the routing and memories that are used to configure the whole architecture. We thus studied how emerging technologies can improve the configuration memory nodes, as well as the routing resources (Chap. 3). Several memory and configurable routing nodes have been proposed using 3-D based technologies. These circuits are expected to improve the global FPGA architecture. These considerations have been addressed by benchmarking a structurally enhanced FPGA (Chap. 4). The benchmarking is performed by using the traditional FPGA flow and by tuning the architectural description.

P.-E. Gaillardon et al., *Disruptive Logic Architectures and Technologies*,
DOI: 10.1007/978-1-4614-3058-2_7,

While improvements of memories and routing resources appeared as the most convenient way to optimize the FPGA structure, we also assessed the logic block circuits (Chap. 5). In fact, computation blocks in standard FPGA systems occupy <20% of the total area. We thus investigated new computation cell paradigms that to enable the exploration of new architectural organizations in which the predominance of the logic blocks is increased. Several technologies were investigated to compact the elementary logic block, leading to ultra-fine grain logic. The standard reconfigurable template was enhanced to take into account the novel ultra-fine grain paradigm. We provided a specific organization for ultra-fine grain logic FPGAs, based on matrices of cells interconnected by fixed and incomplete topologies (Chap. 6). In order to evaluate this new scheme, a specific packing tool was developed and linked to the standard benchmarking flow (Chap. 6). The ultra-fine grain architecture was therefore studied, in terms of topology selection and sizing, and was compared to its equivalent CMOS circuit (Chap. 6).

7.2 Contributions to Methodology and Tools

To allow fast and low-cost assessment of emerging technologies, we developed a new evaluation methodology. In fact, we defined a standard and generic architectural template that was intended to be efficiently realized by disruptive technologies. Thus, it became possible to provide a full set of evaluation tools that allow the evaluation of a wide range of technologies for a broad range of application contexts (Chap. 1). To do so, we proposed a complete tool flow, able to instantiate our generic template customized with several technology parameters (Chaps. 4 and 6). Based on the conventional flow established for the exploration of FPGA architectures, several tools were connected together to allow the evaluation of (i) standard FPGAs, (ii) reconfigurable circuits with enhanced routing, (iii) reconfigurable circuits with gate-based computation blocks (instead of LUTs) and (iv) reconfigurable circuits with a specific architectural organization. In this final context, a specific packing tool was developed in order to allow regular arrangements of logic cells with special interconnection topologies (Chap. 6). This tool (called MPack) is fully compatible with the standard tools and has demonstrated performance in terms of computation time and mapping results compatible with its design exploration purpose.

7.3 Contributions to Memories and Routing Resources for FPGA

In FPGAs, routing resources and memories occupy more than 80% of the chip area. Memories are distributed all over the circuit and are used to program the computational logic blocks and the active routing elements. Memories are traditionally implemented using SRAMs, which are costly in terms of area. Routing resources are used to create paths between logic blocks. The introduction of many active elements through the logic paths has two obstacles. Firstly, they are highly area consuming since they must be built in front-end silicon. This obviously also has an impact on interconnect size, since the lines must constantly connect back-end to front-end active layers and so on. Secondly, the introduction of active devices into the logic data path tends to globally decrease the performance. In order to assess these questions, we explored the interest of three 3-D based technologies (Chap. 3): resistive phase-change memories, monolithic 3-D integration process and vertical NWFET.

Resistive memories are non-volatile passive devices that can be programmed between two stable resistive states. These devices are compatible with the back-end and can be embedded within the interconnection layers. A configuration memory node was proposed, based on resistive memories arranged in a voltage divider, in order to store intrinsically a logic voltage level. The node improves its equivalent non-volatile flash counterpart by 1.5x in area and 16.6x in write time, while it improves on an emerging magnetic memory based equivalent by 3.8x in area. Resistive memories are also remarkable for their low on-resistance (up to 50 Ω, with respect to 9.1 kΩ for minimum-sized CMOS 65-nm transistors). Thus, it is of high interest to introduce the device directly into the logic data paths, to yield high-performance switches. It is thus possible to create routing elements that combine pass-gate and configuration memory into a unique two-terminal resistive node. We demonstrated that this node leads to an improvement in area of 3.4x compared to flash and 14x compared to magnetic memories.

The monolithic 3-D integration process aims to stack several layers of active silicon with high via density. This appears promising with respect to the previous technology that stacks only passive devices. We assessed the performance of FPGA elementary building blocks realized in a full 3-D scheme. Monolithic 3-D technology exhibits high alignment accuracy. This makes it possible to envisage splitting between the configuration memory and the data path logic. In standard 3-D techniques, this is not possible due to the large number of interconnections that are required between the two levels. We implemented a simple 2-bit LUT element and various cross point architectures. We showed that the LUT2 can be improved by 2.03x in area compared to an equivalent 2-D circuit. It has also been shown that the intrinsic delay is improved by 1.62x, the load factor by 6.11x and the average power by 2.02x. This tendency is also observed with the routing cross points where gains of 2.93x in area, 3.1x in intrinsic delay, 1.07x in load factor and 2.1x in average power were demonstrated.

More prospective vertical NWFET technology enables transistors to be entirely implemented between two metal layers. This opens the way towards large high-performance transistors with small footprint impact, due to the vertical orientation of the active volume. This technology also enables the realization of complete logic functions within the metal layers. In particular, we proposed a configurable via that switches on under a Boolean (single- or multiple-input) condition. It is also possible to realize complex logic gates, and so to place signal buffers, logic functions and configuration memories above the initial circuit. To evaluate this disruptive technology, we proposed a methodology based on TCAD simulations. After modeling the devices physically, transient simulations were carried out to extract the performance metrics of simple circuits. We demonstrated that such a technology yields an improvement of 31.2x in area, 2.5x in intrinsic delay and 14.5x in leakage power with respect to the equivalent CMOS gate.

7.4 Contributions to Logic Blocks for FPGA

The largest part of the FPGA architecture is occupied by "peripheral circuitries". In an incremental design approach, it is reasonable to focus on this part. Nevertheless, new computational block architectures should lead to the emergence of new system architectures, and potentially to a reduction of the imbalance between peripheral and logic circuitry resources. In this context, we drastically reduced the size of elementary computational blocks (Chap. 5), and defined the obtained logic as ultra-fine grain logic. Two principal technological paradigms have been explored: the use of enhanced-functionality devices (i.e. devices with innovative functionalities with the same dimensions) and the use of density-increased devices (i.e. more devices in the same space). Regarding the technological obstacles for each technology, we proposed credible technological assumptions for all the expected processes.

For the functionality-enhanced device, we propose to use DG-CNFET technology as case study. A DG-CNFET can be configured to one of three different states (*n*, *p* and *off*) by changing its back-gate voltage. This device enabled the design of a compact dynamic logic cell able to perform 14 Boolean functions with only seven transistors. We showed that such a cell improves area by 3.1x with respect to the equivalent CMOS counterpart due to the increase in functionality, while keeping a power-delay product almost constant with comparisons to a prospective CMOS technology (objectively extrapolated with a methodology based on the ITRS approach and figures). We then provide a generalization of the use of control polarity-gate voltages on dynamic logic cells. In particular, the back-gate is used to select the polarity of the transistor and to control the evaluation of the different stages. We demonstrated a potential gain up to 50% in area compared to standard dynamic MOS logic.

Two density-increased devices process were subsequently explored. Emerging technologies have opened the way towards the use of active devices in a high-density crossbar structure. We proposed the use of a sublithographic crossbar of nanowires to realize an ultra-fine grain logic cell. While such an organization has already been envisaged to build complete and complex systems, we used it to implement small circuits only. The scheme was simplified, in order to manage the connection requirements between the micro and nano scales. In particular, we added internal inverters directly within the structure, instead of propagating inverted signals through the chip. We demonstrated that such a scheme would lead to an improvement of 4.1x in area and of 4.6x in intrinsic delay. This is due to the very small dimensions achieved by the device integration well below lithographic dimensions. Nevertheless, we should note that such a technology is highly controversial regarding the fabrication process. In fact, sublithographic dimensions often require complex alignments of bottom-up fabricated devices. Thus, we proposed a crossbar process flow derived from an industrial and lithographic FDSOI technology, with higher feasibility than the previous approach. We devised the technology guidelines and the layout structure for the logic cells using the crossbar arrangement. Considering all the parasitics, we then performed technological optimization in order to find the best process conditions. We finally demonstrated that the technology yields an area improvement of 6x and a power consumption improvement of 1.48x compared to the CMOS counterpart.

7.5 Contributions to Architectures

As previously presented, the architecture could be improved in two ways: an improvement of routing resources (Chap. 4) and an improvement of the logic blocks that lead to a new architectural organization (Chap. 6).

Improvement of the architecture in terms of memory and routing resources has been done using the various technologies introduced. For each technology explored, a different FPGA organization has been proposed. Thanks to the traditional FPGA benchmarking flow and the tuning of architectural parameters, we showed that it is possible to improve the area, compared to the conventional FPGA scheme, by 46%. This value is obtained for the vertical NWFET technology. This technology is the best in terms of routing circuit area improvement. This obviously impacts the architecture accordingly. Others technologies led to a gain of 21% in area for monolithic 3-D and 13% for phase-change memories. Regarding critical path delay, a maximum gain (up to 44%) was observed for phase-change memories, which is obviously due to the low on-resistance of the technology (i.e. good switch performances). Vertical NWFET technology also led to good results with a gain of 42%. Monolithic 3-D improved the delay by 22%, due to the good electrical properties of the FDSOI technology.

In terms of architectural organization, we described a new scheme suited to computation at ultra-fine grain (Chap. 6). In particular, we explored the use of compact logic cells (Chap. 5). While it is not possible to use a standard FPGA organization due to a large interconnection overload, we proposed the use of matrices of logic cells. These matrices are expected to perform combinational computation. To prevent interconnection complexity, we further proposed the use of a layered interconnect, based on fixed and incomplete topologies. These arrangements were named MClusters, and were introduced in the standard FPGA scheme instead of LUTs. Using the specific benchmarking tools that we developed, we studied the topology impact on the performance of MClusters. We found that the best topology is the Modified Omega, thanks to its properties of interconnect shuffling. We then performed an evaluation of the complete FPGA architecture, and we demonstrated that the best MCluster sizing was obtained for matrices of 3 by 3 logic cells. Such a scheme yielded an improvement of 46% in the total area for the FPGA and a global improvement of the critical path delay repartition.

Index

P.-E. Gaillardon et al., *Disruptive Logic Architectures and Technologies*,
DOI: 10.1007/978-1-4614-3058-2,

L

M